Principles of statistical decision making

MACMILLAN DECISION SERIES

under the editorship of James L. McKenney

Principles of statistical decision making

Dennis J. Aigner

UNIVERSITY OF WISCONSIN

THE MACMILLAN COMPANY, NEW YORK

COLLIER-MACMILLAN LIMITED, LONDON

First Printing

Library of Congress catalog card number: 68-10238

The Macmillan Company, New York

Collier-Macmillan Canada, Ltd., Toronto, Ontario

Printed in the United States of America

To My Parents

HERBERT AND DELLA AIGNER

Foreword

The array of concepts and techniques available to the managerial decision maker is expanding at an accelerating rate. This expansion seems to be caused by the increasingly complex problems the manager must cope with and also by the availability of trained researchers. The control of present-day integrated, international businesses warrants the most sophisticated managerial system feasible. Meeting the challenge of improved control systems in recent years, the large body of researchers has produced a rich spectrum of viable concepts to aid the managerial process. Computer-based information systems of modern business have also contributed to both the development and the economic feasibility of such concepts. The computer with access to an appropriate data base allows the researcher to probe realistic problems and to develop useful solutions. This same computer enables the manager to implement these sophisticated tools with a basic understanding of the approach without the tedium of repetitive data manipulation. Recent computer developments make the combination of powerful concepts with the computer not only feasible but the best economical solution as well.

Unfortunately, one thrust of these recent innovations has created another dilemma for the decision maker: Which approach to select for a given problem. Each managerial concept must be judged according to its appropriateness to the problem at hand, as it requires not only a definition of assumptions and the nature of the important relationships in accordance with reality but also that these assumptions be codified in consonance with the precepts of the concept. Thus, a significant decision for the manager is the selection of the approach to use when faced with a managerial problem. Increasingly, because of his knowledge of the environment and accessibility of the techniques, the manager is called upon to have an understanding of the potential of available concepts and the ingredients necessary to their application.

This series is an attempt to aid the manager's understanding of possible approaches by providing a range of pertinent concepts in a concise and accessible form. The prime criterion of topic selection is usefulness to the manager.

The main elements of usefulness are (1) providing a conceptual framework for approaching a class of problems and (2) providing an analytic technique to aid in the solution of a well-defined problem. Each volume in the series focuses on one topic to afford a comprehensive view of the subject as it relates to the manager.

James L. McKenney

Preface

The purpose of this book is to enumerate the "principles" of statistical decision making or inference. That is to say, by no means do we intend to exhaust the many and varied applications of statistical methods to inference problems. But all such applications possess certain commonalities: assumptions, techniques, and the like. Indeed, they all fit into a basic logical framework or methodology, and it is with this methodology that we will be concerned.

The development of the subject matter proceeds toward the fundamental ideas of statistical inference, as contained in Chapters 5, 6, and 7. To this end the reader is led through the necessary prerequisites and in addition some topics of related interest. Assumed is a familiarity with calculus (a first course) and algebra and a level of general mathematical sophistication sufficient to allow symbols to ease the presentation rather than obscure the ideas.

For the student taking a course in statistics this volume is intended to provide an elaboration of the methodology of inference and, moreover, to present a logical bridge between the classical and modern methods of decision making under uncertainty in as simple a form as possible. This latter purpose has (to the dismay of myself, at least) been ignored in the numerous textbooks in beginning statistics at all levels of presentation. The present text contains no problems, primarily because of its supplementary character.

To a casual reader with the necessary background the book is self-contained. We hope that its pages will provide a motivation to explore the area of statistical decision making (and statistics in general) further. With this anticipation, the following books are recommended:

Nonrigorous presentation, classical orientation:

Guthrie, H., *Statistical Methods in Economics* (Irwin, 1966).
Yamane, T., *Statistics, An Introductory Analysis* (Harper & Row, 1964).

Nonrigorous presentation, modern orientation:

Schlaifer, R., *Introduction to Statistics for Business Decisions* (McGraw-Hill, 1961).

Mathematical presentation, varying degrees of rigor, classical orientation:

Freund, J. E., *Mathematical Statistics* (Prentice-Hall, 1962).
Hoel, P. G., *Introduction to Mathematical Statistics*, 3rd ed. (Wiley, 1962).

Mood, A., and Graybill, F., *Introduction to the Theory of Statistics*, 2nd ed. (McGraw-Hill, 1963).
Neyman, J., *First Course in Probability and Statistics* (Holt, 1950).
Wolf, F., *Elements of Probability and Statistics* (McGraw-Hill, 1962).

Mathematical presentation, modern and classical orientation:

Chernoff, H., and Moses, L., *Elementary Decision Theory* (Wiley, 1959).
Clelland, R. C., et al., *Basic Statistics with Business Applications* (Wiley, 1966).

For a collection of classical tests, a good beginning reference is

Guenther, W. C., *Concepts of Statistical Inference* (McGraw-Hill, 1965).

This list is by no means an exhaustive treatment of the current collection of textbooks in statistics, but should serve to introduce the reader to the subject matter at any predescribed level.

The book begins with a *Reader's Digest* version of itself, and proceeds to topics such as the calculus of probability, frequency distributions, etc., in anticipation of the final chapters. Hopefully, in light of its limited scope, this development serves to build the reader to a proper level of knowledgeable familiarity with the concepts which form the logical framework of statistical inference.

The author wishes to acknowledge the assistance of T. A. Yancey of the University of Illinois in reading the entire manuscript and contributing substantive comment, and the students of its Graduate School of Business Administration, who served yeoman's duty as guinea pigs during the development of the material.

Dennis J. Aigner

Contents

1

Introduction to statistical decision making

The bailiwick of statistics is the class of problems in decision making or inference involving uncertain events or happenings, being, in fact, the applied arm of what is commonly referred to as probability theory—a collection of definitions, postulates, and theorems about uncertain events. And just as our algebra of real numbers yields various formulas for making calculations with them, we shall find it helpful to deduce certain calculation rules about probabilities, the quantitative representations of the "odds" or "chances" that uncertain events will occur. To this end, and for purposes of introducing the concepts and orientation of statistical decision making, we shall proceed from an empirical definition of probability.

1. An experimental definition of probability

Event space is a term denoting the collection of possible outcomes of what we shall call a random experiment. And by *random experiment* we shall mean one or more repetitions of a procedure which has a number of possible outcomes, where only one of the possible outcomes may occur each time the procedure is implemented. Given this definition, then, since the event space is such a collection and every possible outcome of a single repetition of the procedure is a member of it, the event space is termed *exhaustive*. Also, in any repetition of the procedure, by definition only one of the members of the event space can actually occur; the members themselves are called *elementary events*, with the property that elementary events are *mutually exclusive*.

One very simple example of such a procedure is the tossing of a die, where the event space consists of the six faces of the die (the elementary events). The six elementary events are indeed mutually exclusive; only one may turn up on a given toss.

A more commonly used term for the execution of the "procedure" we refer to in a random experiment is *trial*. When we describe such an experiment, then, we may refer to it as consisting of some arbitrary number of trials. For the above example, if a single trial consists of tossing a die once, one of the many possible random experiments we may conceive of would be tossing the die five times.

The primary characteristic of such an experiment, as contrasted to laboratory experiments in physics, for example, is that while we also require reproducibility of this random experiment, its outcome is not governed by so-called natural "laws" but by a *probability law*. To be precise, if we repeated our basic procedure (such as the die toss)—as the physicist would a set of steps in his laboratory procedure—an arbitrary number of times and recorded the number of trials which resulted in each of the elementary events, these numbers being called *frequencies*, we would produce a *frequency table* for the experiment. If we divide each of these frequencies by the total number of trials which comprises our experiment, the resulting quotients, called *relative frequencies*, would exhibit the following properties, owing to the nature of the experiment as defined: Each of the relative frequencies is a positive number in the interval (0, 1), and their sum is precisely 1.

The term *probability* refers to that number to which a given relative frequency tends as the number of trials is increased indefinitely. For example, a "true" die, i.e., one that is perfectly balanced, should exhibit relative frequencies for its six faces which each tend toward $\frac{1}{6}$ as the die is tossed indefinitely. We would then describe a perfectly balanced die as one where the probability that any one of its six possible faces turns up on a given toss is $\frac{1}{6}$.

This last statement contains a subtle but very important point to grasp. It is that while probability is defined via a random experiment with an infinite number of trials, the concept is applied with reference to but *one* of the trials. A statement to the effect that "the probability of a six appearing on a roll of a single die is $\frac{1}{6}$" *means* if we rolled the die an infinite number of times the relative frequency of the elementary event "six turns up" tends toward $\frac{1}{6}$.

We must be careful, however, to understand that such a probability statement may be viewed in two rather different ways: If the statement is made *after* a great many rolls of the die are observed, then it is, in essence, an objective statement about an observed phenomenon.† If, on the other hand, the statement is made *before* the die is thrown, it is subjective in nature, and forms a hypothesis which may be refuted by experimentation. This distinction forms the fundamental difference between probability theory and statistics as two disciplines dealing with uncertain events. Probability theory concerns itself with calculations based on probability assumptions about uncertain events, while statistics deals primarily with testing ex ante propositions about the probabilities of these same events by actual observation.

2. Probability theory vs. statistics—an example‡

To elucidate the previous statement, suppose we are considering placing a bet in a simple game of chance, where a die is thrown four times by a professional gambler. We may bet on or against the appearance of at least one "six" in the four throws with a winning payoff equal to the bet.

The probability problem involved in the decision is to calculate the probability of the event's occurrence before the die has been thrown. Obviously such a calculation rests on certain assumptions about the game itself, primarily a supposition concerning the "fairness" of the die. Being rather familiar with such calculations, we conclude that if the die is "fair" we should bet on the occurrence of this event, since its probability is .518. Given this situation, *and playing the game a sufficient number of times*, we should expect to win some money by betting on the event's occurrence.

The relevant statistical question, on the other hand, relates to the validity of the assumption made about the fairness of the die. Considering

† "In essence," because in practice we can never really establish a probability. But we all may be able to agree on an operational definition. Suppose, for example, in 1000 rolls of a die the relative frequency of the event "six turns up" is .166; in 10,000 it is .1667; in 100,000, .16667, etc. We may agree to say this relative frequency *is* $\frac{1}{6}$ if it agrees to four places, etc.; obviously we shall have to adopt such a rule if we are to make such a statement.

‡ The foundation of organization underlying the next four sections is due to the influence of J. Neyman. The particular pedagogical method of outlining the essentials of a complete course in an introductory lecture or two, particularly in a statistics course, has been successful for the present author, at least. Cf. Neyman's book, *First Course in Probability and Statistics* (Holt, 1950), Chap. 1.

the occupation of the person tossing the die, questioning this assumption would appear quite reasonable. In addition, the die may not be intentionally biased but still not meet the fairness assumption, because of some irregularity in its manufacture. Clearly what we require to say anything about the actual situation that faces us is some additional information about the die in question.

Classical statistics requires that such information consist of observations taken on the experiment, in this case the game. And such observations must be, in fact, capable of description by a probability law. In essence this is the definition of a *probability sample*, a subject we shall devote some space to in Chapter 5. For our use in this introduction, however, we need no more formal definition than this.

Let us, for illustrative purposes, consider as our sample information the observation of two plays of the game, after which we must make a decision as to a bet on or against the event "at least one six." Also, at this point, let us assume that it costs nothing to obtain this information; i.e., our gambling friend will allow us to observe the die without betting for two games.† How should the required decision be made? For example, if both sample games result in the occurrence of at least one six, should we then commence to bet on this event for the subsequent game? Intuitively, if the die is biased against the event in question, such an actuality will be less likely than if the die is fair or biased in the other direction. That is to say, the sample result will be more consistent with a hypothesis of fairness or bias in favor of the event than bias against its occurrence. But, of course, since the sample consists merely of a *finite* number of observations on the random experiment (in this case, two) it cannot purport to identify exactly the true probabilities governing the game. Thus such a decision may be incorrect because it is based on a *decision rule* which cannot discriminate perfectly between the various underlying probabilities which *could have* caused such a sample outcome and that probability set which is the true one. It is precisely to this problem of inference that statistics addresses itself explicitly through the relation of the true parameters to sample "estimates" of them, or implicitly by examining the probability characteristics of decision rules based on sample results.

3. Statistical inference I: estimates and parameters

The explicit role of statistics, alluded to above, involves such inferences as can be made from the sample to the *parameters* of the random experiment in question. In the case at hand, what information does the sample result yield about the true probabilities—the parameters—governing the game? One rather intuitive *estimator* of the probability of the event "at least one

† Usually obtaining sample information will not be free; hence the selection of a "best" sample size, if it exists, is an important matter. We shall discuss this point later in this chapter and elaborate upon it in Chapters 6 and 7.

six in four throws of a die" from the sample would be the relative frequency of the event in the sample, since by definition the limiting value of the relative frequency (as the number of sample observations in increased indefinitely) is the probability in question.

For example, if in our sample of two games the event did not occur, our sample *estimate* of the event's probability would be zero. If it occurred once our estimate would be $\frac{1}{2}$, and if it happened that both sample games resulted in at least one six the estimate would be 1. Abstracting for the moment from the implications of any *particular* result on the decision problem, can we say anything about the appropriateness of the *estimator* itself (i.e., the way in which we utilize sample information) as an approximator for the parameter in question?

One such comment has already been made: If the *sample size* (number of observations) is increased indefinitely, the estimator (relative frequency of the event in the sample) will—indeed, must—become essentially indistinguishable from the true parameter: the probability of the event. And we can, in this case, obtain as precise an estimate of this probability as we like merely by obtaining more observations. But in general such precision may not be secured without some cost; and usually such cost (e.g., sampling cost) will increase in proportion to increases in precision. Thus in most realistic applications there is some tradeoff between precision in an estimate and the costs necessary to obtain given precision; therefore, there exists a basis for some optimization of the two.

How "precise" must a given estimate of some parameter be? And, indeed, what do we mean operationally by "precision"? An answer to the latter question requires much additional preparation and is treated formally in Chapter 5. For our present purposes we shall ask only for an intuitive understanding of the concept in the context of our particular estimate-parameter example, as discussed above. Precision of an estimate will be taken to mean how "close" it is in value to the parameter. The operational problem derives from the unhappy circumstance that a parameter is usually an unknown—else why would we be interested, as a practical matter, in obtaining an estimate of it?

To elaborate on the practical aspects of estimation and "precision," however, we can accomplish much from an informal approach. One prerequisite to proceeding further, however, is to note that to speak of how precise an estimate must be only makes sense if we can relate it to the consequences of being *imprecise* in the given problem of inference. To be specific, we must be able to state, quantitatively, the ramifications of using any estimate as a surrogate for the parameter of interest.

4. Statistical inference II: decision rules

Our present example of a decision problem under uncertainty provides a relatively good basis for discussing precision of an estimate and the effects of precision on decisions (inferences) about a parameter (or the parameters) of some random experiment.

To begin let us attempt to trace, as completely as possible at this introductory stage, the ramifications of using some *fixed* rule of behavior, determined ex ante, to decide between a bet on or against the event "at least one six" in four throws of a single die. Let us also suppose, for simplicity, that these are the only courses of action open and also that the game is to be played a great many times.† Further, consider the consequences of adopting the specific decision rule:

> $\mathbf{R}_1$. Bet against the occurrence of "at least one six in four throws" if, in a sample of two observations of the game, the event is *not observed.* Otherwise, bet on the event's occurrence, i.e., when it is observed either once or twice in two games.

Intuitively, if the parameter governing the game, i.e., the probability of a six turning up on a given throw of the die, is biased upward or is even that value one would associate with a "fair" die, we would like to adopt a rule which would lead to betting on the event in question. Similarly, the rule should concomitantly be able to discover when the probability of a six turning up on a single throw is biased downward enough to make betting against the event "at least one six in four throws" desirable (when this probability is .159 or less). It should be intuitively obvious from our previous discussion that that R_1 cannot possibly meet these criteria simultaneously. In fact, we cannot devise any such rule for a finite sample size. The reason derives from the empirical definition of probability. Only when our sample is infinitely large —when the relative frequency of the event becomes its probability precisely— will we be able to devise such a rule.

Short of this unimaginably large number of sample observations we can, however, obtain a rule which guarantees that we shall bet against the event when our parameter, call it p, is $< .159$, and, therefore, meets one of our criteria for an "optimum" decision rule. All we need do is to *always* bet against the event, no matter what the sample information is. Thus, if the parameter p is $< .159$, we shall most certainly make the correct decision. Of course, there is one very large "string" attached to the use of this rule: when p is $> .159$, we *always* make the wrong decision.‡ On the other hand, were we to bet on the event independent of what is observed in the sample we would eliminate completely the above problem: When $p > .159$ we shall always be making the right choice of action. Needless to say, the price we pay for this accuracy is an incorrect action whenever $p < .159$. As long as both these errors "cost" us something, it is generally

† Both assumptions will be either modified or discarded later.

‡ It is assumed here that when $p = .159$ we would be indifferent as to a bet on or against the event.

true that the decision rule we do adopt will contain inherent errors of both kinds. Indeed, this may be the case for any rule constructed for a finite sample in our example where the costs of the two sorts of errors are positive.

The selection of a rule proceeds, therefore, from a number of possibilities, including the two extreme cases and encompassing all possible rules in between. For our sample of two games, the number of alternative rules is at least manageable.

To elaborate on the problem of choice of decision rules from among the alternatives available for a given number of sample observations,† we shall use R_1 for the example and consider the characteristics of this fixed rule with respect to the possible errors discussed above.

Suppose p is, in fact, equal to $\frac{1}{6}$. In two plays of the game, our sample set of observations, there are calculable probabilities (chances) that any one of the three (mutually exclusive) events, no occurrence, one occurrence, two occurrences, of "at least one six" *does* occur. These values can be calculated as .233, .499, and .268, respectively. Now since R_1 calls for a given action in correspondence with these same events, we may easily speak of the "probability" that R_1 will cause us to follow one of the two alternative actions, bet against the occurrence of "at least one six in four throws" or for it, *given* $p = \frac{1}{6}$. These probabilities are .233 and .767, the latter being the sum of two of the probabilities given before, for the events one and two occurrences of the event in the sample.

Since we have supposed p to be $\frac{1}{6}$, R_1 may be evaluated on this conditional basis. We should like to bet on the occurrence of "at least one six" in this case, but our calculations demonstrate that the probability is .233 that R_1 will cause us to bet against the event, an incorrect decision. R_1 may be evaluated in analogous fashion for all values of $p < .159$, .159 being the "breakeven" value, where we would, in a lengthy series of plays, expect to win and lose an equal number of games no matter which event we bet on. In fact, R_1 causes us to err in exactly the way described above for all values of p in the range $.159 < p < 1$. However, when $p < .159$, R_1 behaves in a somewhat different fashion. Suppose $p = \frac{1}{10}$. Then the probability of at least one six in four throws is .344, and we certainly would like to bet against the event in this case. For the sample of two, if $p = \frac{1}{10}$, then R_1 correctly causes us to bet against the event whenever no occurrences are observed. The calculated probability for this occurrence is .430. Thus the probability is .570 that R_1 causes us to incorrectly bet on the event when $p = \frac{1}{10}$. A complete set of such error-probability computations for all $0 < p < 1$ is shown graphically in Figure 1-1. From mere observation, and assuming both types of errors are equally distasteful, it appears that R_1 is a good rule to use if p actually has some value in the range $.159 < p < 1$.

† We shall see in Chapter 7 how both the best rule and optimum sample size may be determined simultaneously.

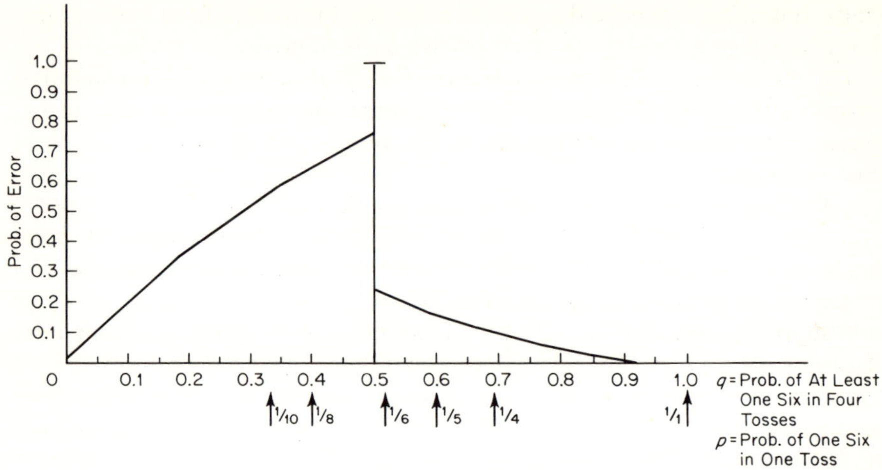

Figure 1-1. Error characteristic for rule R_1, $n = 2$.

But what of alternative rules which may be considered in competition with R_1? In fact, if we allow some such "alternative" rules to be discarded on the grounds of their intuitive unreasonableness,† we have but three remaining competitive rules, two of which are our extreme cases, discussed previously. The other, say R_2, directs a bet against the event whenever it is observed one or zero times in the sample of two plays. Otherwise, i.e., when the event occurs twice, R_2 causes us to bet on the event. Following procedures and interpretations laid down previously, Figure 1-2 is constructed, where rules R_3 and R_4 correspond to the rules causing bets always on and always against the event, respectively.

From the graph it would seem that no "optimum" rule from among these four alternatives exists. Indeed, some *judgment* must be made concerning the characteristics of each rule in conjunction with an evaluation of the decision-maker's losses from committing the errors involved in choice under uncertainty. This is a consequence of the fact that no one rule possesses smallest probability of error over the complete range of p.

As we shall see in the development that follows, the nonexistence of a "best" rule in the above sense is an impasse often encountered in classical statistics. The modern contribution follows as a technique of judging competitive decision rules based on additional information about the problem, in the form of judgments about the likely (range of) values for the parameter

† For example, the rule "bet on the event when no occurrences are observed in the sample" is an "unreasonable" alternative. We use this appeal to intuition here only for the sake of convenience. In principle *all* alternative rules should be considered objectively. Indeed sometimes our intuition will not be correct. The reader is referred to Section 8, Chapter 2, for a good example of the perils of using intuition as a substitute for the actual calculation of probabilities.

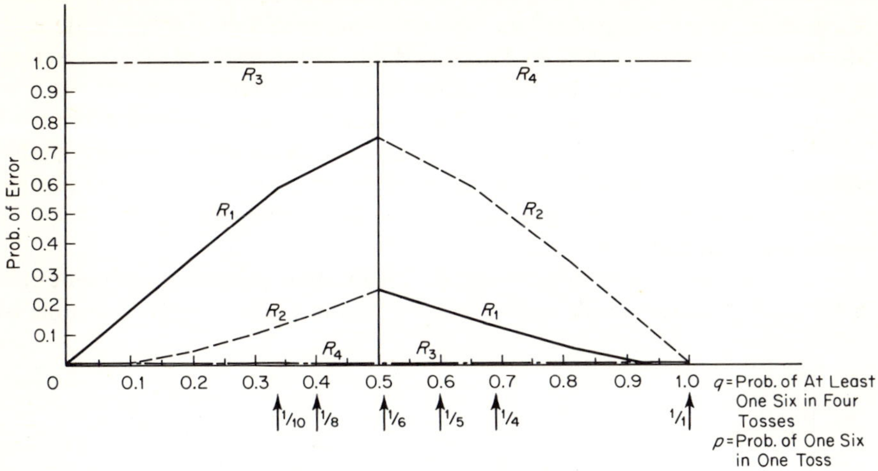

Figure 1-2. Decision rules: error characteristics for $n = 2$.

in question. At the basis of the modern contribution is a redefinition of the word *parameter* in its classical sense, that of a fixed characteristic (magnitude) of some phenomenon subject to description by a probability law.

5. Further considerations for the selection of decision rules

Our introduction has so far been very limited in scope, in that the example used conforms poorly to the complex decision problems of the real world. Some problems will perhaps fit even our very simple model quite well; most others will not. The intent in this chapter and, indeed, in the entire book is to present principles of application and not a catalogue of cases which must be used like a series of physical conversion factors. So, although every possible complexity (and/or variation in decision models under uncertainty) will not be explicitly treated, we shall indicate the ingredients and outline the recipe for their inclusion.

For example, in Section 4 we considered alternative decision rules whose behavior was dependent on the value of the underlying parameter p, but of one *form* only. That is, another set of alternative rules for making the same decision is available based on the number of times a six appears in eight throws of a single die. Clearly we are not altering the available sample information, merely the way in which it is utilized. For an exhaustive treatment of the selection problem, error-probability functions for the seven possible rules in this set should have been considered as well. The method to be followed in arriving at their evaluation is precisely that used before. Only the way in which the probabilistic behavior of the rules is calculated is changed.

Of the other limitations of this simple example, the matter of a fixed, predetermined sample size deserves elaboration in this introduction. As was pointed out previously in reference to the ideas of estimation and precision of estimates, there is in general an increase in the precision of an estimate with additional sample information. Translated into terms relevant to the selection of a decision rule, as sample size is increased indefinitely we should observe the graph of error probabilities (cf. Figure 1-2) to collapse toward a family of functions where *one* decision rule becomes able to dictate a correct choice *always*. This statement is basically an inductive interpretation, *via the specific example*, of the definition of probability applied to the statistical problem of choice under uncertainty.

As an illustration of this conjecture, results analogous to those contained in Figure 1-2 (for a sample size of 2) are given in Figure 1-3 for a sample of

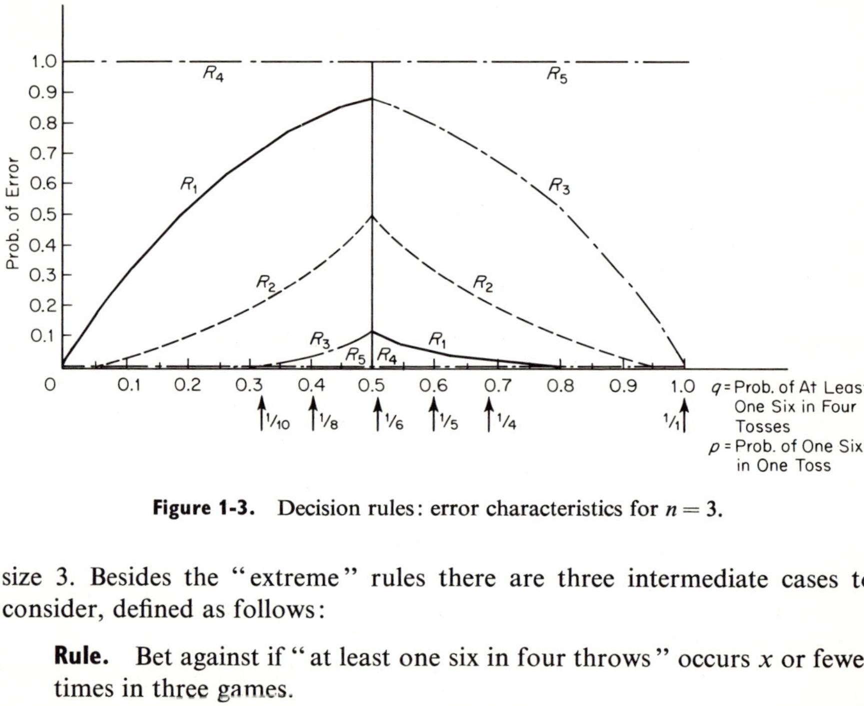

Figure 1-3. Decision rules: error characteristics for $n = 3$.

size 3. Besides the "extreme" rules there are three intermediate cases to consider, defined as follows:

Rule. Bet against if "at least one six in four throws" occurs x or fewer times in three games.

Rule	x
R_1	0
R_2	1
R_3	2

In this graph the extreme cases are represented as R_4: always bet on the event and R_5: always bet against the event.

The important comparison to be made between these two figures relates to the effects of an increase in sample size of one unit. We note in Figure 1-3 that R_1 possesses more extreme error probability values for the $n = 3$ case than it did for $n = 2$ (cf. Figure 1-2). The newly introduced R_3 similarly has low error probabilities when $0 \leq p \leq .159$ and high probabilities when $.159 \leq p \leq 1$, so no easy choice between these rule alternatives is possible. Indeed, R_2 is not a "best" rule either, although it has an error characteristic which lies between the other functions over the complete range of p. If we compare R_2 in Figure 1-3 to rules R_1 and R_2 of Figure 1-2, however, we see the effect of the additional sample information clearly in that by a choice of R_2 in Figure 1-3 we limit the maximum error probability to .500 over the entire range of p. Both R_1 and R_2 of Figure 1-2 have maximum errors of .750. Further increases in sample size will produce a similar reduction in the maximum error probability over the entire range of p via *one* of the alternative rules available at each sample size. This is not to say that such rules are "optimum" in any cost sense. It merely points out the fact that as sample size increases the general level of error probabilities decreases.

As we shall see later, such a statement as this is tantamount to suggesting increased precision of an estimate as sample size increases. In the limit of sample size, errors in the estimate and error probabilities for one decision rule must approach zero in the sense decribed above.

In an ex ante sense, the decision model should also include an evaluation of how large the sample "should" be. To present a method for obtaining a simultaneous solution of both problems, sample size and optimum rule, is the goal of this book. To accomplish the goal requires a development of material which continues through the following five chapters and culminates in Chapter 7.

2

Elements of the calculus of probabilities

The rudiments of probability theory may be approached in a variety of ways, from the very rigorous to the highly intuitive. Something positive may be said for both approaches, although any depth in the subject must depend on a rigorous, mathematical presentation. Our treatment here will follow a position somewhere between the extremes, in that the calculus of probability is developed intuitively with only suggestions as to formal proofs. As will also be true with the rest of the material we cover, whenever the rigor is easily implemented it will be presented.

1. Sample space—elementary events

Sample space and event space are terms we shall use interchangeably to denote the collection of elementary events which form the set of possible outcomes of a random experiment. We begin by considering such sets of events as finite, and represent them usually in terms of a collection of "points" in some coordinate system. For example, the simplest such sample space—for reasons we shall enumerate presently—might be represented by two points on a line. If we code the possible outcomes of the toss of a coin as zero for "tails" and 1 for "heads," the sample space would look as shown in Figure 2-1.

Figure 2-1. Sample space for the toss of a coin.

The characteristic condition placed on such a discrete sample space is that it does, indeed, consist of separate points; that is, if one of the elementary events is known to have occurred, the probability that any other elementary event has also occurred is zero. This condition, which may apply to other than the elementary events, is that elementary events are defined to be *mutually exclusive*. They are also *exhaustive*: The set of elementary events is the sample space.

Let us represent the probability associated with each elementary event, say E_i, as p_i. Then from the definition of probability the following statements hold

$$0 < p_i < 1, \tag{1.1}$$

$$\sum_{i=1}^{n} p_i = 1, \tag{1.2}$$

where the number of possible outcomes of the random experiment, i.e., the elementary events, is n. The first condition merely implies that each elementary event must be a "possible" outcome and that a random experiment must consist of at least two elementary events, otherwise one event would be certain to occur. The second condition follows directly from the definition of probability.

2. Compound events—the addition theorem for mutually exclusive events

For a working example of a sample space we shall use the set of elementary events generated by tossing a coin twice. The elementary events for this random experiment may be easily listed as follows:

1st toss	2nd toss
H	H
H	T
T	H
T	T

And, if we adopt the same coding convention used in Figure 2-1, the sample space may be drawn in a two-dimensional coordinate system, letting the results of the first toss be graphed on the vertical, or y, axis, and the results of the second toss be plotted on the horizontal, or x, axis. This sample space is shown as Figure 2-2, with the coordinates written in (y, x) order. Applying

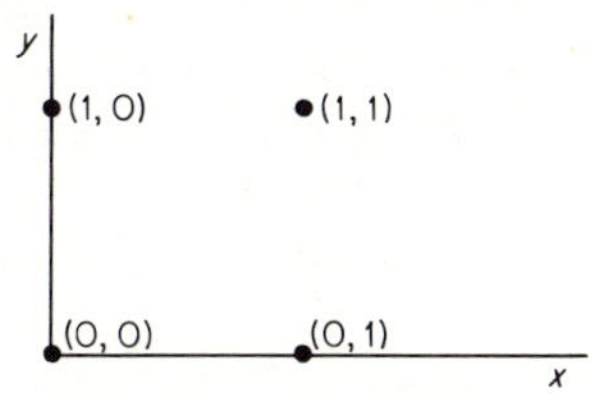

Figure 2-2. Sample space for the tossing of a coin twice.

our adopted notation, the correspondence of the coordinates to the (E_i, p_i) representation is given as follows (numbering arbitrary):

Event	Notation	Probability
(0, 0)	E_1	p_1
(0, 1)	E_2	p_2
(1, 0)	E_3	p_3
(1, 1)	E_4	p_4

We discussed previously the general relationship between probability and statistics, the major difference being that for probability theory the probability structure (set of p_i's) is assumed for the particular model at hand, whereas for statistics such an assumption is usually under direct investigation as a hypothesis to be confronted with empirical evidence. For this example, then, the set of four p_i's is assumed known for purposes of illustration. If the coin is "fair," each elementary event should receive an a priori probability of $\frac{1}{4}$.†

† If this assignment of probability weights is not obvious now it shall become so later, as we may derive this random experiment heuristically from a single coin toss.

Now we turn to so-called *compound* events, that is, events which are said to occur whenever one of a number of elementary events occurs. The event "one head in two tosses" is such a compound event, since it occurs when either of the elementary events E_2 or E_3 occurs. Since E_2 and E_3 are elementary events, they are mutually exclusive; it is obvious that the probability weight which should be attached to the event "one head in two tosses" is simply the sum of the probabilities p_2 and p_3. Under our assigned probability structure this event has a probability of occurring of $\frac{1}{2}$.

In fact, any compound event may conceivably be treated in terms of its constituent elementary events. Its probability may be obtained as a simple sum (for the finite sample space) of the appropriate p_i's. For use in the remainder of this chapter it will be convenient to adopt a general notation for writing this probability relation between the elementary events and any other event defined on the sample space: If A is some event in the sample space S, then the probability that A occurs is the probability that any one of the elementary events in A occurs. Letting the set of elementary events $\{E_i\}$ contained in and forming A be denoted by $E_i \in A$, where the symbol "$\in$" means "belongs to," then the probability attached to A is

$$P(A) = \sum_{E_i \in A} p_i. \tag{2.1}$$

At this point it is also convenient to introduce the notion of *complementary* events. Given the above definition of an arbitrary event A in terms of elementary events, we shall call the event defined by the elementary events *not* in A "A complement" or $\bar{A}$. Thus

$$P(\bar{A}) = \sum_{E_i \in \bar{A}} p_i = 1 - \sum_{E_i \in A} p_i, \tag{2.2}$$

by definition of the sample space. A and $\bar{A}$ are two mutually exclusive events which "taken together" in probability terms define the whole space S. In set-theory terminology, we are equating S to the *union* ($\cup$) of A and $\bar{A}$: the set of points which is contained in *either* A *or* $\bar{A}$ (or both). Since A and $\bar{A}$ are mutually exclusive, they, by definition, have no common points. Thus we have the relationship

$$P(S) = P(A) + P(\bar{A}) = \sum_{E_i \in S} p_i = 1. \tag{2.3}$$

This relationship, when written for arbitrary events C and B *which are mutually exclusive*, is a special case of the general addition theorem for probabilities. We present it as

Rule 1. *Addition—mutually exclusive events.* For mutually exclusive events C and B,

$$P(C \cup B) = P(C) + P(B) = \sum_{E_i \in C \cup B} p_i. \tag{2.4}$$

As an example of the use of rule 1, consider the event A, "at least one head in two tosses." This event may be viewed first as a subset of the original sample space: the points E_2, E_3, and E_4. By definition,

$$P(A) = \sum_{E_i \in A} p_i = p_2 + p_3 + p_4 = \tfrac{3}{4}.$$

This event might be given in terms of two other events, B: "one head in two tosses" and C: "heads on both tosses." Rule 1 applied to event B yields $P(B) = p_2 + p_3 = \frac{1}{2}$. C is just the elementary event E_4 and so $P(C) = p_4 = \frac{1}{4}$. Thus $P(A) = P(B \cup C) = P(B) + P(C)$, since B and C contain no common points.

3. General addition rule

Suppose we consider the "union" probability, say $P(A \cup B)$, where the events A and B are defined on our example sample space of two coin tosses as the events "heads on the first toss" and "heads on the second toss," respectively. Considering the union of these events, i.e., the event "heads on the first or second toss," it should be clear that, under our assumed probability structure, $P(A \cup B) = \frac{3}{4}$. The reasoning is simple: What is required is the sum

$$P(A \cup B) = \sum_{E_i \in A \cup B} p_i = p_2 + p_3 + p_4 = \tfrac{3}{4}. \tag{3.1}$$

Indeed, any event may be treated in terms of its constituent elementary events, conceptually, at least. That is, it may be that the sample space cannot be easily enumerated or identified. Such would be the case were we to ask for the probability that an arbitrary American citizen has red hair, given that we have an operational definition of "red hair." Although the sample space in this case is finite, it is impossible, as a practical matter, to identify it for many reasons. The required probability may, however, be calculable via other probabilities, of events in the same sample space. Moreover, and again as a practical matter, even though the sample space may be available, it may be more efficient to calculate some probabilities in terms of others. To thus develop rules of calculating with probabilities would, given any concern with them in the first place, appear a worthwhile effort. In this regard rule 1 may be generalized to include probability calculations for the union of events which are not mutually exclusive.

Considering the calculation of the event described above, where $P(A \cup B)$ was discovered to equal $\frac{3}{4}$, our goal is to relate this probability, if possible, to the probabilities of the individual events, i.e., $P(A)$ and $P(B)$. Writing $P(A)$ and $P(B)$ in terms of their individual sample-space probabilities, it is clear that the correspondence between $P(A \cup B)$, $P(A)$, and $P(B)$ is in the form of a simple algebraic sum with a correction term:

$$P(A \cup B) = \sum_{E_i \in A \cup B} p_i = p_2 + p_3 + p_4 = P(A) + P(B) - p_4 . \tag{3.2}$$

The correction factor corresponds to the probability that *both* events, A and B, occur.

Extending the rule to cover arbitrary events A and B, it is sufficient to note that the correspondence between $P(A \cup B)$ and $[P(A) + P(B)]$ will be exact if an amount equal to the probability of the common points of A and B is subtracted. That is, in general,

$$(A \cup B) = P(A) + P(B) - \sum_{E_i \in A \text{ and } B} p_i . \tag{3.3}$$

In set-algebra terms the event "*A and B*" corresponds to the *intersection* ($\cap$) of A and B, written $A \cap B$: that event which occurs whenever A and B occur together. Obviously there will be no such event unless A and B are defined to contain at least one common point. If no common point exists, A and B are mutually exclusive and the special addition rule applies. The general rule contains the special case since $P(A \cap B) = 0$ when $A \cap B$ contains no points, and is equivalent to (3.3), or:

Rule 2. *General addition of probabilities.* For A and B two events defined on the sample space S,

$$P(A \cup B) = P(A) + P(B) - P(A \cap B). \tag{3.4}$$

Using rule 2 on the example, we have $P(A) = \frac{1}{2}$, $P(B) = \frac{1}{2}$, $P(A \cap B) = p_4 = \frac{1}{4}$, so that $P(A \cup B) = \frac{1}{2} + \frac{1}{2} - \frac{1}{4} = \frac{3}{4}$.

4. Conditional probability and the general multiplication rule

By rule 2 we may determine any $P(A \cup B)$ with knowledge of $P(A)$, $P(B)$, and $P(A \cap B)$. What of further relationships which might assist us by giving $P(A \cap B)$ in terms of $P(A)$ and/or $P(B)$? One such relationship, that of *conditional* probability, is of use in this empirical sense and also has independent importance as a useful concept in much of probability theory and statistics.

The "condition" is that information is available, in an ex post sense, about the experiment, in the form of knowledge that an event, say C, *has occurred.* The event of interest, say D, is then treated in a question of the form, "Given that C *has occurred*, what is $P(D)$?" Clearly from the question the elementary events involved in this conditional probability are those common to D and C, i.e., the events defining the intersection event $D \cap C$, and are therefore of particular relevance to us at this stage of the development of our calculus of probability. It is not true, however, that the probability of this intersection event is the answer to the conditional question.

An example may be helpful in demonstrating this last statement. Suppose heads occurred on the first toss of the coin. What is the probability that heads occurs on the second toss? Since the same coin is assumed to be tossed both times, the probability structure governing the second toss is clearly unaffected by any result of the first toss being known. The answer to this conditional question is, from this reasoning, $\frac{1}{2}$. The combined result of two heads in the two tosses is clearly the intersection of the events A and B of Section 3, "heads on the first toss and heads on the second toss." This event has a probability of occurring of $\frac{1}{4}$. So, although both probability questions, the conditional and the intersection questions, concern the *same* sample-space points, some *modification* of the a priori probability structure for the intersection event is required to treat the conditional question.

What "modification" is required? If the event C, "heads on the first toss," *has occurred*, then any probability calculations concerning other events contained in C must be made *as if* C were a complete sample space in its own right. That is to say, before probabilities of events in C may be obtained, C must meet the conditions imposed on the probabilities of elementary events in any sample space: The probabilities must all lie between zero and 1 and must sum (over all elementary events) to 1 [see (1.1) and (1.2)]. Clearly, since the elementary events in C have assigned probability weights p_i from S, they meet the first condition. However, their sum is $P(C)$, not 1. Therefore, some or all such weights must be inflated by amounts such that the sum of inflated probability weights equals 1.

To be sure, the inflation procedure cannot be arbitrary if we are to preserve the *relative* importances, in probability terms, of the elementary events in C. For example, in S the probability relationship between the events "two heads in two tosses" and "heads on toss 1 and tails on toss 2" is a one-to-one correspondence, $p_4 = p_3 = \frac{1}{4}$. *This relationship must still hold in the restricted sample space*, C, even though the weights p_3 and p_4 are not appropriate to describe the above events when C has already occurred. But from the above reasoning, we can be sure that the "appropriate" probabilities, say π_3 and π_4, differ from their counterparts in S, p_3 and p_4, by the same factor of proportionality.

In this specific case the solution for π_3 and π_4 is easy, since we know they must satisfy $\pi_3 = \pi_4$, and $\pi_3 + \pi_4 = 1$, viewed as if C had already occurred. The solution is obviously $\pi_3 = \pi_4 = \frac{1}{2}$.

In general the required conditions may be stated

$$\pi_i = kp_i \qquad \text{all } E_i \in C, \tag{4.1}$$

$$\sum_{E_i \in C} \pi_i = 1. \tag{4.2}$$

Substituting (4.1) in (4.2), we obtain the result

$$k \sum_{E_i \in C} p_i = 1, \tag{4.3}$$

and then

$$k = \frac{1}{\sum_{E_i \in C} p_i} = \frac{1}{P(C)}, \tag{4.4}$$

by definition. For conditional probability calculations—when C has already occurred—we should first multiply all the probability weights for elementary events in C by k, the reciprocal of $P(C)$. From this new probability structure any conditional probability may be easily written.

For the event under consideration, "D given C," which we write $D \mid C$, the required probability is available by rule 1 as

$$P(D \mid C) = \sum_{E_i \in D \cap C} \pi_i, \tag{4.5}$$

the sum of probabilities π_i for those elementary events of C which are also contained in D, that is, the intersection $C \cap D$. Since $\pi_i = kp_i$, with k given by (4.4), we have

$$P(D \mid C) = \frac{\sum_{E_i \in D \cap C} p_i}{P(C)} = \frac{P(D \cap C)}{P(C)}. \tag{4.6}$$

Therefore, for any event C so that $P(C) \neq 0$, the conditional question may be answered by a ratio of the intersection probability of C with the specific event to the probability of C. Both probabilities in the computation are derived from the original sample space, S. By writing (4.6) in a different fashion, we obtain the general multiplication rule for probabilities.

Rule 3. *General multiplication rule.* For two events C and D in S,

$$P(C \cap D) = P(D \mid C)P(C), \tag{4.7}$$

$$P(C \cap D) = P(C \mid D)P(D). \tag{4.8}$$

Equation (4.8) follows directly from an interchange of events C and D in (4.6).

Considering the example already introduced, let D be the event "heads on the second toss" and let C be the event "heads on the first toss." Then

$P(C) = \frac{1}{2}$ and $P(D \mid C) = \frac{1}{2}$, because knowing the conditional sample space consists of the events E_3 and E_4 implies that the conditional probability for these equiprobable events is $\frac{1}{2}$. Using rule 3 we obtain $P(D \cap C) = (\frac{1}{2})(\frac{1}{2}) = \frac{1}{4}$, which is correct from the original event space, S. Thus we can work back and forth from conditional probability to intersection probability easily, an important result, as we shall see by Illustration 1 of Section 8.

The concept of conditional probability is also important from the standpoint of ex post or a posteriori probability calculations. For instance, Bayes' theorem, which is concerned with general conditional probability calculations, is an important rule in the "modern" approach to statistical decision making of Chapter 7; it is developed in a later section.

5. Statistical independence

The example chosen in Section 4 as an illustration of the ideas of conditional probability happens also to serve as an example of events which are independent in probability, or *statistically independent*. Verbally, two events are said to be statistically independent if information as to one event's occurrence has no effect on the probability structure of the other event. In symbols this statement may be translated, in terms of events C and D, as

$$P(D \mid C) = P(D) \text{ implies independence.} \tag{5.1}$$

Event D, therefore, bears the same relation in probability to the restricted sample space C as it does to the original sample space, S, when D and C are "independent." It must be clearly understood that the independence of two events does not mean the events are mutually exclusive. Indeed, if two events are mutually exclusive, knowledge that one of them has occurred implies that the other could not possibly have occurred; i.e., if the events R and T are mutually exclusive, $P(R \mid T) = P(T \mid R) = 0$. The notion of statistical independence depends on the intersection of the two events containing at least one elementary event.

The formal definition of independence derives from (5.1) applied to rule 3:

Rule 4. *Statistical independence.* Two events, D and C, are said to be *statistically independent* if

$$P(D \cap C) = P(D)P(C). \tag{5.2}$$

Condition (5.1) is clearly equivalent to (5.2) as a definition of independence and must also imply that the other possible conditional probability is equal to its appropriate "unconditional" probability; i.e., $P(C \mid D) = P(C)$ when $P(D \mid C) = P(D)$, and vice versa.

For the example, the event "heads on the first toss" is independent of the event "heads on the second toss" by rule 4, since $P(C \cap D) = \frac{1}{4} = P(C)P(D) = (\frac{1}{2})(\frac{1}{2})$, or by the fact that $P(D \mid C) = P(D) = \frac{1}{2}$. Like the concept of conditional probability, statistical independence has significance in an empirical sense, as we might calculate an intersection probability by rule 4 from reasoning that two events must be independent a priori, and in the sense that two arbitrary events may be identified as independent if they meet the condition (5.2) [or, equivalently, (5.1)].

6. Extension of the addition and multiplication rules to multiple events

As most models of human behavior involve numerous "variables" in attempting to account for such behavior, they are multidimensional in nature. Likewise we shall have occasion to calculate probabilities for combinations of more than two random events. The rules for calculation already presented may easily be extended to cover such cases, at least conceptually. For our purposes the rules for addition and multiplication involving three events will be given with brief suggestions toward the development of the most general cases.

Rule 5. *Extended general addition rule.* For events A, B, and C in a sample space S, the probability of their mutual union is given by

$$P(A \cup B \cup C) = P(A) + P(B) + P(C) - P(A \cap B) - P(A \cap C) - P(C \cap B) + P(A \cap B \cap C). \quad (6.1)$$

This identity is intuitively obvious if we appeal to a visual interpretation of the required probability. Consider the events A, B, and C of rule 5 as they appear in Figure 2-3. Were we to delineate the probability of the event

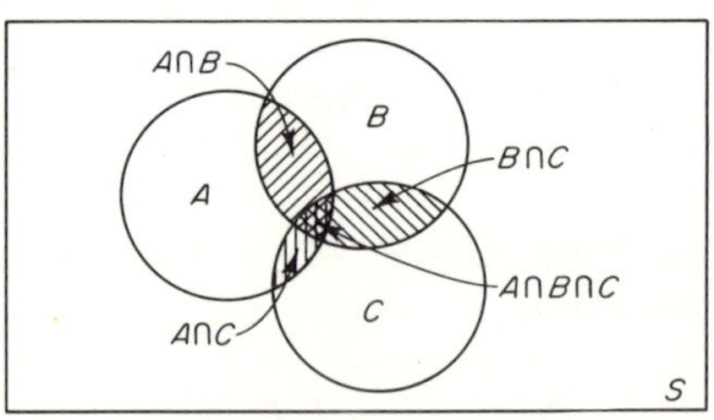

Figure 2-3.

$A \cup B \cup C$ (heavy border in Figure 2-3) by its elementary events in terms of familiar quantities such as the probabilities of A, B, and C, and their mutual intersections, we would find that the probability in question may indeed be written in these terms with a correction factor consisting of the probability associated with that subset of elementary events common to A, B, and C jointly. Thus we need only define the intersection event $A \cap B \cap C$ to produce the rule.

Addition of probabilities for more than three events proceeds easily by letting event C in rule 5 be equivalent to the union event $(G \cup F)$. Then, for example, $A \cap C = A \cap (G \cup F) = (A \cap G) \cup (A \cap F)$, so that $P(A \cap C) = P(A \cap G) + P(A \cap F) - P[(A \cap G) \cap (A \cap F)]$ by rule 2. The last set is the same as $A \cap G \cap F$. An analogous representation in $P(B \cap C)$ follows directly. For the joint intersection term, $A \cap B \cap C = A \cap B \cap (G \cup F) = A \cap (B \cap G) \cup (B \cap F) = (A \cap B \cap G) \cup (A \cap B \cap F)$.† We may then use rule 2 to calculate this probability.

The rule for multiple intersections likewise follows inductively from the analogous rule for two events. Using rule 3 but substituting the event $(A \cap B)$ for D in (4.8), we get

$$P(C \cap A \cap B) = P(C \mid A \cap B)P(A \cap B). \tag{6.2}$$

The last probability also may be represented in conditional terms as $P(A \mid B)P(B)$ or $P(B \mid A)P(A)$. The result appears as

Rule 6. *Extended multiplication rule*

$$P(C \cap A \cap B) = P(C \mid A \cap B)P(A \mid B)P(B). \tag{6.3}$$

Such a "chain" of conditional probabilities may clearly be constructed for an arbitrary number of events merely by repeating the indicated substitution procedure and applying it in (6.3).

This rule may be interpreted for the case of statistical independence by applying rule 4 first to the right-hand-most part of equation (6.3) and then again, to obtain

$$P(C \cap A \cap B) = P(C)P(B)P(A), \tag{6.4}$$

if A, B, and C are statistically independent events.

Rule 6 will be illustrated in one of the examples of Section 8. The identity for statistical independence of multiple events bears directly on much material in statistics. Its primary importance, from the standpoint of the present text, is contained in sampling theory and estimation (cf. Chapter 5).

† The reader should convince himself of these identities by following the indicated steps in a diagram such as Figure 2-3 if the distributive law for sets is not familiar.

7. Bayes' theorem

The concept of conditional probability and its relation to the intersection event allows an interesting and useful interpretation in terms of "cause-and-effect" relationships among uncertain events. Such an interpretation depends on the condition that the sample space S may be described in terms of the union of a number of mutually exclusive "regions." For instance, take the sample space of Section 2, our coin-tossing example, and suppose that one head in the two tosses was observed. The "bayesian" question is one of cause and effect; specifically, *given* this outcome, what is the probability that it was due to the sequence heads-tails rather than tails-heads on the first and second tosses?

One may correctly question the usefulness of asking such a question based on this simple example, because to make a statement such as "one head occurred in two tosses" someone must have observed either the heads-tails or tails-heads sequence. For the time being, let us establish a temporary rationale for considering such a problem as if we were intending a bet with the someone who did the observing as to which "cause" was responsible for the "effect."

Since the question calls for the calculation of a conditional probability, we can use rule 3 to discover the correct answer. The correspondence between the symbolic events of rule 3 (4.6) and the actual events of this problem are as follows: Event C, the "condition," is equivalent to the "effect," that one head in two tosses was observed; event D, representing one of the "causes," we shall let represent the heads-tails sequence. Then the fully verbalized question is, "What is the probability that heads occurred on the first toss and tails occurred on the second, given that one head in two tosses *was* observed?" It seems intuitively obvious that the answer should be $\frac{1}{2}$, since only two possible mutually exclusive "causes" exist and were equally probable in the original sample space.

To use the calculation rule $P(D \mid C) = P(D \cap C)/P(C)$, we can easily identify $P(C) = \frac{1}{2}$, the probability that one heads in two tosses is observed in S. $P(D \cap C)$ is the probability that one head in two tosses is observed *and* that heads-tails is the observed sequence. This intersection event is just the event D itself, so $P(D \cap C) = P(D) = \frac{1}{4}$, and $P(D \mid C) = (\frac{1}{4})/(\frac{1}{2}) = \frac{1}{2}$. From this simple case no need for a general calculation rule for such conditional problems is apparent.

Consider as another "cause-effect" problem the following situation, the "boxes and balls" of Figure 2-4. A box is selected at random with equal

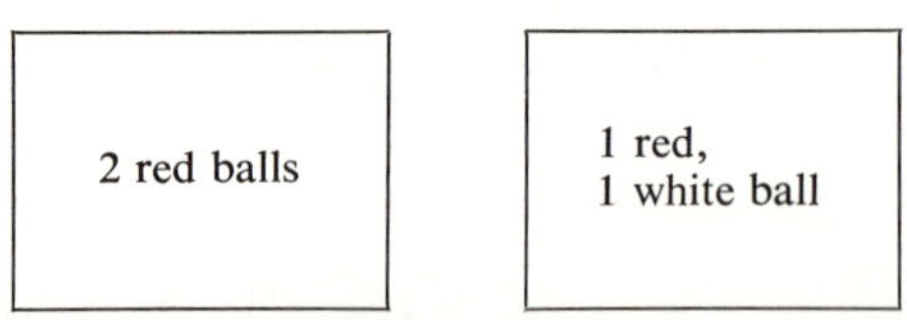

Figure 2-4. Data for boxes-and-balls example.

probability; i.e., either box has probability $\frac{1}{2}$ of being selected. If a ball is then selected from the chosen box at random (again, equal probability of selection among balls in a box), what is the probability that if a red ball is selected it was chosen from box 1? This is a conditional question, box 1 given a red ball.

For the sake of analysis let us define the events

A_1: box 1 chosen
A_2: box 2 chosen
B_1: red ball chosen
B_2: white ball chosen

The sample space of the experiment—of selecting a box and then a ball according to the assigned probabilities for such selections—consists of certain intersections of these events, as shown in Figure 2-5. The question asked

Event	**Probability of Event**
$A_1 \cap B_1'$	$\frac{1}{4}$
$A_1 \cap B_1''$	$\frac{1}{4}$
$A_2 \cap B_1$	$\frac{1}{4}$
$A_2 \cap B_2$	$\frac{1}{4}$

Figure 2-5. Sample space for the boxes-and-balls problem.

concerns the conditional probability $P(A_1 \mid B_1)$; thus its answer may be obtained by the general procedure of restricting the sample space to B_1, properly inflating the probabilities of elementary events in B_1, and finally, calculating $P(A_1 \mid B_1)$. Since the four events have equal probability in S, the original sample space, the three of these events in B_1 must preserve this relationship. Thus the restricted sample space is that shown in Figure 2-6. The required

$P(E_i)$	**E_i**
$\frac{1}{3}$	$A_1 \cap B_1'$
$\frac{1}{3}$	$A_1 \cap B_1''$
$\frac{1}{3}$	$A_2 \cap B_1$

Figure 2-6. Restricted sample space for boxes-and-balls problem.

probability is the probability of the union of $A_1 \cap B_1'$ and $A_1 \cap B_1''$ in this restricted sample space. This probability is $\frac{2}{3}$.

Using a more heuristic approach, rule 3 applied to this example gives

$$P(A_1 \mid B_1) = \frac{P(A_1 \cap B_1)}{P(B_1)}. \tag{7.1}$$

To determine $P(A_1 \cap B_1)$ we make use of the symmetry of the intersection–conditional probability relationship [e.g., (4.7) and (4.8)] to find

$$P(A_1 \cap B_1) = P(B_1 \mid A_1)P(A_1). \tag{7.2}$$

$P(B_1 \mid A_1)$ is just the probability that a red ball is selected given that the first box is drawn, and equals 1, since only red balls are contained in box 1. $P(A_1)$, likewise, is readily available, being the probability that box 1 is selected to begin with, and equals $\frac{1}{2}$ by specification. Then $P(A_1 \cap B_1) = P(B_1 \mid A_1)P(A_1) = (1)(\frac{1}{2}) = \frac{1}{2}$ for the numerator of (7.1).

Event B_1, drawing a red ball (unconditionally), may occur in two ways as it is obtained from box 1 or box 2. Thus $B_1 = (A_1 \cap B_1) \cup (A_2 \cap B_1)$. Since A_1 and A_2 are mutually exclusive "regions" of the sample space S, the union of their intersections with a given event, say B_1, must yield exactly the set of points in B_1. These intersections are mutually exclusive events, so we may write

$$P(B_1) = P(A_1 \cap B_1) + P(A_2 \cap B_1), \tag{7.3}$$

from rule 1. The first of these probabilities was already computed as $\frac{1}{2}$, and the second may be obtained in a similar way, i.e., by writing $P(A_2 \cap B_1) = P(B_1 \mid A_2)P(A_2)$. Again, these latter probabilities are available directly, as $\frac{1}{2}$ and $\frac{1}{2}$, respectively. Therefore, $P(B_1) = \frac{1}{2} + \frac{1}{4} = \frac{3}{4}$, and as a final result we have

$$P(A_1 \mid B_1) = \frac{P(A_1 \cap B_1)}{P(B_1)} = \frac{\frac{1}{2}}{\frac{3}{4}} = \frac{2}{3}. \tag{7.4}$$

The "novel" use made above of the symmetry relation of conditional and joint (intersection) probability may be extended easily to a formal rule. Let the mutually exclusive regions alluded to previously be k in number, and represented by $H_1, \ldots, H_k$. Then any event in the sample space $S = H_1 \cup \cdots \cup H_k$, say A, may be represented in terms of mutually exclusive intersections of A with the H_j's, as in Figure 2-7.

The "bayesian" problem is to determine the probability that an arbitrary H_r has occurred given that A has occurred, which is given by rule 3 as

$$P(H_r \mid A) = \frac{P(H_r \cap A)}{P(A)}. \tag{7.5}$$

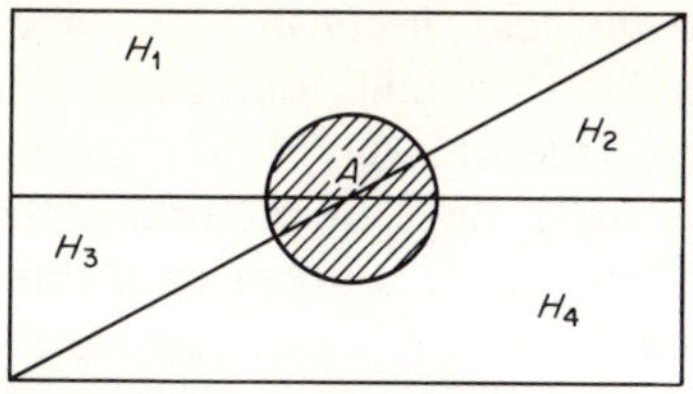

Figure 2-7. Partitioning of S into mutually exclusive regions.

Using the symmetry relationships of rule 3 we may write $P(H_r \cap A) = P(A \mid H_r)P(H_r)$. Also, because the H_j's are mutually exclusive, $P(A)$ may be represented through rule 1 in terms of the union of the mutually exclusive intersections of A with the H_j's. That is,

$$P(A) = P(H_1 \cap A) + \cdots + P(H_k \cap A) = \sum_{j=1}^{k} P(H_j \cap A). \tag{7.6}$$

These intersections, in turn, can be subjected to the symmetry argument so that

$$\sum_{j=1}^{k} P(H_j \cap A) = \sum_{j=1}^{k} P(A \mid H_j)P(H_j). \tag{7.7}$$

Equation (7.5) may thus be rewritten so all required probabilities are in the natural cause-effect chronology, as

Rule 7. *Bayes' Theorem.* For $H_1, \ldots, H_k$, mutually exclusive and exhaustive regions of S, and an event A in S,

$$P(H_r \mid A) = \frac{P(A \mid H_r)P(H_r)}{\sum_{j=1}^{k} P(A \mid H_j)P(H_j)}. \tag{7.8}$$

An additional example of the use of rule 7 is given as Illustration 2 of Section 8.

8. Probability and intuition: two illustrations

In this final section of our introduction to the calculus of probability the major point made is that although intuition may play an important role in guiding actual decisions under uncertainty, it is many times a poor judge of the actual probability structure involved. In support of this appeal for a more "intelligent intuition" two illustrations are provided. The first requires knowledge of the multiple union and/or multiple intersection rules, and the second is a further application of rule 7.

Illustration 1. *The birthday problem.*† In a group of individuals of arbitrary size, n, what is the probability that there is in the group at least one duplication of birthdays (month, day)? My own experience is that the probability calculation usually made by the uninitiated is to look at the fraction $n/365$ (disregarding leap years), and consider this to be the probability of the event occurring. Such reasoning, when applied to the same problem in a betting situation, could make statistics professors financially independent in a short time, since students who make the above computation seem perfectly willing to bet and *give* odds of even money to 10 or 15 to 1 when n is of moderate classroom size, say 25 to 30.

Of course, the actual probabilities involved are somewhat more difficult to compute than this simple ratio, but are nonetheless straightforward.

Attacking the problem head-on, the probability in question may be viewed as one of multiple union. What is required, heuristically, is the computation first for the probability of *exactly* one duplication, then for *exactly* two, etc. The probability of the union of these events yields the required probability. The amount of computation is certainly large, however, in comparison to an alternative calculation, suggested by a knowledge of the concept of complementary events (cf. Section 2). That is, the complement of the event "at least one duplication" is the event "no duplications in a group of size n." These two events make up the entire sample space and by definition, then, P (at least one duplication in n) $= 1 - P$ (no duplications in n).

The "no duplication" event is a multiple intersection: What is required is the probability that the first individual have a brithday different from the other $(n - 1)$ individuals, who, in turn must have birthdays different from each other's, etc. We may easily compute this probability for any n by using induction on some simple cases.

Take, for example, the two-individual case: The probability of no duplication of birthdays is equal to the probability that individual 1 have a birthday different from individual 2, i.e., that his birthday be one of the 364 of 365 remaining days. The required probability is simply 364/365 that there will be *no* duplication.

Extending to three individuals, we use the multiple intersection rule (rule 3) to obtain the result. If $A \mid B$ is the event "individual 3 has a birthday different from individuals 2 and 1 (given *their* different birthdays)," then the event *no* duplication is equivalent to the event $A \cap B$, that 3 has a birthday different than 1 and 2 *and* that 1 and 2 *have* different birthdays. Thus the required probability may be written

$$P(A \cap B) = P(A \mid B)P(B), \tag{8.1}$$

† Also well documented and illustrated in the Time-Life Science Series, Vol. I, Mathematics.

from rule 3. The last probability, $P(B)$, equals 364/365, as above. $P(A \mid B) = 363/365$, since 3 must have one of the remaining 363 birthdays different from 1 and 2. Thus $P(A \cap B) = (364)(363)/365^2$.

These results may be generalized to calculate the probability of *no* duplications from the formula

$$P(\text{no duplications in } n) = \frac{(365)(364)\cdots(365-n+1)}{365^n}. \tag{8.2}$$

The probability of at least one duplication is available, as mentioned before, as 1 minus this probability.

The results of actual calculations are certainly inconsistent with the intuition described previously. For example, when $n = 23$, the probability of at least one duplication is just over $\frac{1}{2}$. As n increases from 23, progress of the required probability toward 1 is rapid, as seen in Table 2-1.

Table 2-1. Probability structure of the birthday problem

n	Probability of occurrence	*n*	Probability of occurrence
1	0	**24**	.53834427
2	.00273973	**25**	.56869972
3	.00820417	**26**	.59824084
4	.01635592	**27**	.62685930
5	.02713558	**28**	.65446149
6	.04046249	**29**	.68096855
7	.05623571	**30**	.70631625
8	.07433530	**31**	.73045464
9	.09462384	**32**	.75334753
10	.11694818	**33**	.77497186
11	.14114138	**34**	.79531687
12	.16702479	**35**	.81438324
13	.19441028	**36**	.83218211
14	.22310252	**37**	.84873401
15	.25290133	**38**	.86406782
16	.28360402	**39**	.87821966
17	.31500768	**40**	.89123181
18	.34691143	**50**	.97037358
19	.37911854	**60**	.99412266
20	.41143840	**70**	.99915958
21	.44368835	**80**	.99991433
22	.47569532	**90**	.99999385
23	.50729725	**100**	.99999969

Illustration 2. *Coins and boxes.* There are three boxes, each with two drawers. The drawers contain coins as follows: In the first box each drawer contains a gold coin; in the second box each drawer contains a silver coin;

and in the last box a gold and a silver coin are contained in its two drawers, respectively. If equal probability selection methods are used to select a box, then a drawer, what is the probability that if a gold coin is selected it was obtained from box 3? The answer most usually offered to the above problem (yet for apparently different reasons) is $\frac{1}{2}$. A popular line of reasoning is based on the fact that only two of the boxes contain gold coins. Hence, if a gold coin is selected, it could have come from one of two equally likely boxes, with probability $\frac{1}{2}$.

Proper reasoning follows lines similar to the analysis of the balls-and-boxes problem of Section 7, and the correct answer is available directly from an application of rule 7, Bayes' theorem. If we let the events B_1, B_2, and B_3 represent the selection of the three boxes, respectively, and also let G be the event "a gold coin is drawn," then symbolically rule 7 gives the required probability, $P(B_3 \mid G)$, as

$$P(B_3 \mid G) = \frac{P(G \mid B_3)P(B_3)}{\sum_{i=1}^{3} P(G \mid B_i)P(B_i)}. \tag{8.3}$$

The probability of drawing a particular box is specified as $\frac{1}{3}$, and given any box, each drawer "receives" a probability of selection of $\frac{1}{2}$.† With these probabilities accounted for, (8.3) becomes

$$P(B_3 \mid G) = \frac{(\frac{1}{2})(\frac{1}{3})}{(1)(\frac{1}{3}) + (0)(\frac{1}{3}) + (\frac{1}{2})(\frac{1}{3})} = \frac{1}{3}. \tag{8.4}$$

† For instance, by flipping a fair coin to determine which drawer to select. To select a box we might use a fair die, assigning arbitrarily two of the six faces to each box.

Counting problems in probability calculations

Many interesting probability problems involve the counting of objects—in sequences, combinations, etc. For example, before the rules of probability calculus may be applied to questions relating to one's possible holding of cards in bridge, poker, etc., we must first be able to enumerate possible configurations of hands. As will be seen in the following chapters, such counting problems are also important as they apply to further development.

The two basic counting concepts involved in looking at a given number of objects as they may be arranged into subgroups are the notions of a sequenced subgroup, called a *permutation* of the original set of objects, and a nonordered subgroup of these same objects, called a *combination* of them.

Suppose there are n objects to be sequentially considered in groups of three. The total number of 3-permutations of n (≥ 3) may be obtained by enumeration as follows: Choose one object. The number of separate ordered groups of two from n is then $n(n-1)$, since we may make the initial choice from n objects, the second choice from $(n-1)$ and, finally, the third choice from $(n-2)$. Thus the total number of 3-permutations is equal to $(n)(n-1)(n-2)$. Applying induction we obtain the number of r-permutations from $n(n \geq r)$ as

$$P_r^n = (n)(n-1) \cdots (n-r+1). \tag{2A.1}$$

A convenient computational version of (2A.1) is available by converting it into terms of factorials. That is,

$$P_r^n = n(n-1) \cdots (n-r+1) = \frac{n!}{(n-r)!}, \tag{2A.2}$$

where $n! = n(n-1)(n-2) \cdots (2)(1)$, etc.

For example, the number of 3-permutations from 5 equals $5 \times 4 \times 3 = 60$ from (2A.1). Equivalently, $(5 \times 4 \times 3 \times 2 \times 1)/(2 \times 1) = 60$. By definition $0! = 1$, so when $n = r$ we find there are exactly $n!$ n-permutations of n elements.

The number of unordered groups, called combinations, of size r from a number of objects n ($n \geq r$) can be obtained easily by noting from (2A.2) that there are exactly $k!$ permutations (i.e., ordered sets) of any arbitrary k elements from n. Thus the total number of r-permutations of n elements must differ by a factor of $r!$ from the number of r-combinations of these elements. Specifically, the number of r-combinations of n elements is given by†

$$\binom{n}{r} = P_r^n/r! = \frac{n!}{r!(n-r)!}. \tag{2A.3}$$

Consider the problem of computing the probability of drawing four aces in a five-card poker hand. The total number of different five-card hands possible is

† The familiar binomial coefficient of algebra. We shall use the notation $\binom{n}{r}$ throughout this book although C_r^n is also common.

given by (2A.3) as $\binom{52}{5}$. We then compare this number to the number of different (unordered) ways four aces in five cards may come about. The hand must contain all four aces in conjunction with any one of the other 48 nonaces. The required count of such possible hands is simply 48.

This result may also be obtained through computation with (2A.3) as the number of 4-combinations of 4 cards (the aces) in conjunction with (times) the number of 1-combinations of the remaining 48 nonaces. The required probability is the ratio of the number of ways the specific hand may be obtained divided by the total number of five-card hands; i.e.,

$$P(4 \text{ aces in } 5 \text{ cards}) = \frac{\binom{4}{4}\binom{48}{1}}{\binom{52}{5}} = \frac{5!}{(52)(51)(50)(49)}$$

$$\simeq .00002. \tag{2A.4}$$

Such calculations as this bear a close relationship to our rule of computation for joint events (rule 3). As an illustration of this correspondence, consider the calculation of the probability that the first three cards received in a five-card poker hand are aces. This probability must be equivalent to the probability of obtaining three aces in three cards, which is given by

$$P(3 \text{ aces in } 3 \text{ cards}) = \frac{\binom{4}{3}\binom{48}{0}}{\binom{52}{3}} = \frac{4!}{(52)(51)(50)}$$

$$\simeq .0002. \tag{2A.5}$$

This probability can also be obtained through rule 3 directly, as seen in Table 2A-1 and computed as equation (2A.6):

$$P(3 \text{ aces in first } 3 \text{ cards}) = \frac{4}{52}\,\frac{3}{51}\,\frac{2}{50}$$

$$= \frac{(4)(3)(2)}{(52)(51)(50)} = \frac{4!}{(52)(51)(50)}. \tag{2A.6}$$

Table 2A-1.

Event	Probability of event	Type of probability
Ace 1 on draw 1	$\frac{4}{52}$	Unconditional
Ace 2 on draw 2	$\frac{3}{51}$	Conditional on draw 1
Ace 3 on draw 3	$\frac{2}{50}$	Conditional on draw 2

Calculations for particular holdings in any randomly dealt game of cards may be obtained through extensions of the basic method. As a final illustration, what is the probability that a bridge deal will result in four "perfect" hands, where each player holds an entire suit? The probability calculation is obtained via rule 3, where we compute the joint event of four perfect hands by the probability of one such hand (unconditional) with the probability that given the one perfect hand, the next individual receives one also (of a different suite, hopefully), etc. The entire computation is shown as follows:

$$P(\text{four perfect hands}) = \frac{\binom{13}{13}\binom{39}{0}}{\binom{52}{13}} \underset{\substack{\text{conditional} \\ \text{on 1st}}}{\frac{\binom{13}{13}\binom{26}{0}}{\binom{39}{13}}}$$

$$\times \underset{\substack{\text{conditional} \\ \text{on} \\ \text{1st and 2nd}}}{\frac{\binom{13}{13}\binom{13}{0}}{\binom{26}{13}}} \underset{\substack{\text{conditional} \\ \text{on 1st, 2nd,} \\ \text{and 3rd}}}{\frac{\binom{13}{13}\binom{0}{0}}{\binom{13}{13}}}$$

$$= \frac{1}{\binom{52}{13}\binom{39}{13}\binom{26}{13}\binom{13}{13}} \simeq 2 \times 10^{-19}. \tag{2A.7}$$

The denominator of (2A.7) is just the total number of four-hand deals in bridge ($\simeq 5.3 \times 10^{18}$).

3

Probability distributions

At the foundation of statistical inference lie calculations such as those just discussed with reference to random events. However, instead of speaking in terms of events and their associated probabilities in the sample space, the terminology and, to be sure, the concepts, are altered to capture another aspect of random experiments, phenomena, and the like.

Consider as an introductory illustration of the need for additions to our definitions and concepts the following problem of statistical inference: A probability sample of individuals is taken from a list of all personal telephone subscribers in a certain community. Each individual selected is asked to submit information on his family size, income, and occupation, the intention being to obtain insights into some of the attributes of the population of telephone subscribers. The first two such "variables" are quantitative in nature, the latter one is qualitative. All three are clearly defined over the set of individuals in the assumed population, and take on their particular "values" in a one-to-one correspondence with the elements of the population.

Suppose a single individual is selected by a procedure which guarantees all individuals an equal chance of being selected.† Then the population of individuals possesses all the necessary prerequisites to be a valid sample space, with the individuals playing the roles of elementary events. When one individual is selected, each of the aforementioned variables takes on, as a result of the selection, a particular value, real numbers in the case of family size and income, and some occupational classification for the qualitative variable. Viewed ex ante, each variable possesses certain possible values, each defined as a result of a compound event; that is, the variables take on their respective possible values as certain individuals are selected. There may be 500 doctors in the population, so that the variable "occupation" takes on the value "doctor" when any of these 500 persons is selected. Similarly, the other variables may be treated in this fashion, for certain individuals will have equal incomes, family sizes, etc.

Since we are sampling from this population, the probability that a certain individual is selected is related to the probability that the variable "income" takes on a value equal to that individual's income. "Related" since there may be other ways the income variable may take on the same value, i.e., via the *possible* selection of other individuals with the same income. Obviously the probability that the variable income does take on that value, say x, is equal to the sum of probabilities corresponding to the selection of one of the group of individuals having x as their income. In fact, if the population is finite, we can list each value of x with such an associated probability.

The other variables mentioned, although being defined on the same sample space, call for different groupings of individuals, with the result, in general, that their lists of possible values will have associated probabilities which both differ mutually and with the income variable.

It is certainly possible to redefine the suggested variables in terms of new sample spaces, etc., so that no additional conceptualizing is required to make probability computations with them. The fact that each such variable is derived from a common sample space suggests, however, that such computations may be simplified conceptually and/or practically by using this fact in defining such variables generally. Moreover, if the sample space consists of a large number of elementary events, the heuristic approach may be completely

† By a table of random numbers, perhaps.

infeasible as a practical method for manipulating probabilities. Thus additional conceptual work appears to rest on a firm rationale.

1. Random variable and frequency function defined

The essential features of the aforementioned variables, which take on values with certain probabilities, allow them to be described in a functional context as *random variables.* A random variable, then, is a function whose domain is the sample space (i.e., defined over the sample-space "points") and which "maps" the elementary events onto the real line.† Although such a definition describes all random variables as quantitative,‡ it does so without conceptual loss, since any qualitative variable may be arbitrarily coded numerically (e.g., as with the coin-tossing examples of Chapter 2). If the sample space in question contains a finite number of elementary events, any random variable defined on it must possess only a finite number of possible values, and hence is described as *discrete.* On the real line the discrete random variable appears as a set of discernible points.

As the sample space contains greater numbers of elementary events, to a conceptual limit of infinity, random variables defined on such a space may (although not necessarily) also have an infinite number of possible values in their ranges. Such variables are considered *continuous* in their ranges, i.e., on the real line or a segment of it. Clearly it is not possible to define a continuous random variable on a finite sample space, but it is possible to define a discrete one on a nonfinite sample space.

The *frequency function* or probability distribution of a discrete random variable accomplishes the task of collecting those elementary events which yield given values of the random variable and summing their probabilities. Thus the domain of the frequency function is the set of possible values for a discrete random variable, and its range is (0, 1), since its values are probabilities. If a random variable is discrete, a graph of its probability distribution would consist merely of points. For example, Figure 3-1 shows visually the

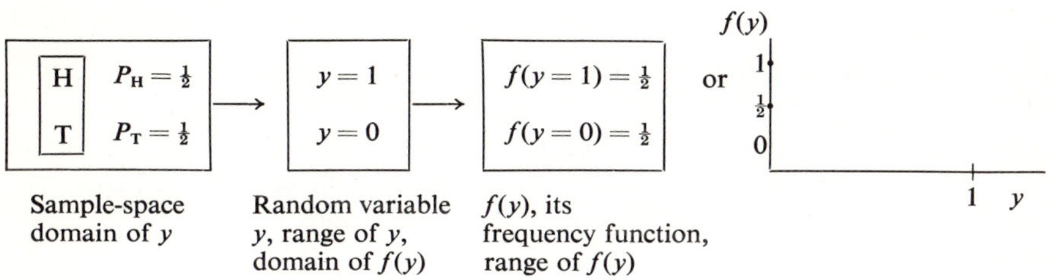

Figure 3-1.

† Or range of real numbers $(-\infty, \infty)$.

‡ To eliminate the problems associated with using qualitative inputs to a "function" and interpreting the resulting "values" of the function.

process of creating and representing a random variable, y; in this case y represents the result of a coin toss.

For the discrete case, any frequency function must meet certain obvious conditions to be a frequency function. The conditions derive directly from the definition of $f(y)$.

For a discrete random variable, y:

$$\text{(a)} \quad 0 \leq f(y) < 1$$

$$\text{(b)} \quad \sum_{\substack{\text{all values}\\ \text{of } y}} f(y) = 1 \tag{1.1}$$

$$\text{(c)} \quad p(y = k) = f(k)$$

In (1.1a), the boundary conditions follow again directly from previous work. If any $f(y) = 0$, then that y value is not "possible"; however, it is sometimes helpful to allow $f(y)$ for some y to be zero if only for brevity of notation. So, although we shall speak only of the "possible" values of random variables, sometimes the mathematical formulation will contain $f(y) = 0$ for some y values.

Continuous random variables differ from discrete ones only in mathematical representation, the major difference deriving from the fact that the frequency function for such a variable is a continuous, differentiable curve.

For a continuous random variable, x:

$$\text{(a)} \quad f(x) \geq 0$$

$$\text{(b)} \quad \int_{-\infty}^{\infty} f(x)\, dx = 1 \tag{1.2}$$

$$\text{(c)} \quad P(a \leq x \leq b) = \int_a^b f(x)dx,$$

where the range of x is assumed to be the real line. Part (c) is a consequence of (b), and relates the fact that the probability in the continuous case can be stated only as a "density" or area under $f(x)$ over a given interval (a, b), $a \neq b$. There is no probability for a given point, x, as can be seen from the following demonstration: Let the interval in question be $(a, a + \Delta a)$, $\Delta a > 0$. Then by (1.2c),

$$P(a \leq x \leq a + \Delta a) = \int_a^{a+\Delta a} f(x)\, dx = [a - (a + \Delta a)]f(\theta), \tag{1.3}$$

for $a \leq \theta \leq a + \Delta a$, from the mean-value theorem of calculus. If we now let

$\Delta a \to 0$, we see that the probability (1.3) goes to zero as $\Delta a \to 0$. The conclusion is that $P(x = a) = 0$, so that a continuous random variable cannot yield a point probability. Of course, the converse is true of discrete random variables.

As an illustration of the above ideas, consider the function $f(x) = ke^{-x}$. First, is $f(x)$ a frequency function? Condition (1.2a) calls for $f(x) \geq 0$, and if $k > 0$ this condition will be met. However, on the unrestricted interval $(-\infty, \infty)$ there is no value of k which will yield condition (1.2b). If the range of x is restricted to nonnegative values only, however, then

$$\int_0^\infty ke^{-x}\,dx = -ke^{-x}\Big|_0^\infty = 1 \tag{1.4}$$

is the condition to be satisfied. It can only be met when $k = 1$, so the valid function is $f(x) = e^{-x}$. Part (c) of (1.2) yields probability calculations for intervals (a, b) on the line $(0, \infty)$; e.g.,

$$P(1 \leq x \leq 2) = \int_1^2 e^{-x}\,dx = e^{-1} - e^{-2} \cong .23. \tag{1.5}$$

Figure 3-2 shows this probability as the appropriate area under the

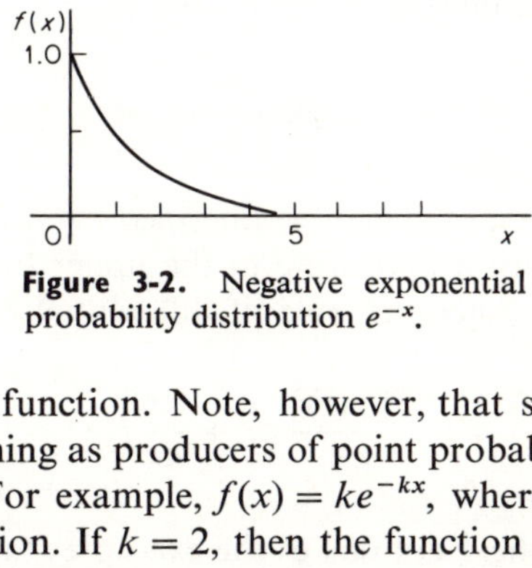

Figure 3-2. Negative exponential probability distribution e^{-x}.

negative exponential function. Note, however, that since continuous distributions have no meaning as producers of point probabilities, the height of an $f(x)$ may exceed 1. For example, $f(x) = ke^{-kx}$, where $k > 0$ and $x \geq 0$ is a valid frequency function. If $k = 2$, then the function yields a value $f(x) = 2$ for $x = 0$. But this number has no independent significance in probability terms. For this same distribution, $P(0 \leq x < 1) = -e^{-2x}\big|_0^1 = 1 - e^{-2} =$ .865, a meaningful probability "density."

2. Distribution functions

Many common frequency distributions are tabled for obvious reasons. To circumvent highly elaborate tables, use is made of the simple fact that for an arbitrary random variable, x, defined on the interval (α, β),

$$P(a \leq x \leq b) = \int_a^b f(x)\,dx = \int_\alpha^b f(x)\,dx - \int_\alpha^a f(x)\,dx, \qquad \{\alpha \leq a,\, b \leq \beta\} \tag{2.1}$$

for continuous distributions, or

$$P(a < x \leq b) = \sum_{k=a}^{b} f(x = k) = \sum_{k=\alpha}^{b} f(x = k) - \sum_{k=\alpha}^{a} f(x = k), \tag{2.2}$$

for the discrete case, if a and b are possible values of x. If not, then the probability statement (2.2) needs modification either in presentation or interpretation.

The cumulant of a given distribution, usually considered from the lower bound (α) of the range of the random variable to some arbitrary point, is called the *distribution function* of the random variable, written $F(x)$. Thus by definition

$$F(x) = \sum_{r=\alpha}^{x} f(r) \qquad \text{if } x \text{ is discrete,}$$

or (2.3)

$$F(x) = \int_{\alpha}^{x} f(r)\, dr \qquad \text{if } x \text{ is continuous.}$$

For the continuous case $F(x)$ bears a significant relationship to $f(x)$, since it is essentially the integral of $f(x)$. That is,

$$\frac{dF(x)}{dx} = f(x) \qquad \text{where } F(x) = \int_{\alpha}^{x} f(r)\, dr. \tag{2.4}$$

Thus the density function of x may be obtained from $F(x)$ as follows: Since we require $dF(x)/dx$ with x appearing in the upper limit explicitly, we may use the "function-of-a-function" rule of differential calculus to obtain

$$\frac{dF(x)}{dx} = \int_{\alpha}^{x} \frac{d[f(r)]}{dx}\, dr + \frac{dx}{dx} f(x) = f(x) \tag{2.5}$$

as required. Applying these notions to the example $f(x) = e^{-x}$, $x \geq 0$,

$$F(x) = \int_{0}^{x} e^{-r}\, dr = 1 - e^{-x}, \tag{2.6}$$

$$\frac{dF(x)}{dx} = (-1)(-e^{-x}) = e^{-x} = f(x). \tag{2.7}$$

Using (2.6) to calculate $P(1 \leq x \leq 2)$, we get

$$P(1 \leq x \leq 2) = (1 - e^{-2}) - (1 - e^{-1}) = e^{-1} - e^{-2} \tag{2.8}$$

as before [cf. equation (1.5)].

3. Joint frequency functions

To complete the analogy to probability calculations for events in terms of frequency functions, we first note that more than one variable may be defined on a given sample space. Thus to speak of a frequency function which yields joint or conditional probabilities is meaningful just as the concept of various events defined over the same sample space has meaning. The "function" of such a frequency function remains to collect the probabilities of elementary events involved in producing the values of random variables jointly or conditionally.

Based on the main example of Chapter 2, we may define the random variables x and y representing the results of the first and second toss as follows:

$$\begin{aligned} x &= 0 \text{ if T on 1st toss} \qquad & y &= 0 \text{ if T on 2nd toss} \\ &= 1 \text{ if H on 1st toss,} & &= 1 \text{ if H on 2nd toss.} \end{aligned} \tag{3.1}$$

$$\begin{aligned} &f(x=0) = \tfrac{1}{2}, & &f(y=0) = \tfrac{1}{2}, \\ &f(x=1) = \tfrac{1}{2}, & &f(y=1) = \tfrac{1}{2}, \\ &\therefore \quad f(x) = \tfrac{1}{2}, \qquad x = 0, 1, & &\therefore \quad f(y) = \tfrac{1}{2}, \qquad y = 0, 1 \end{aligned}$$

and a *joint frequency function*, $f(x, y)$ as

$$f(x, y): \begin{cases} f(x=0, y=0) = \tfrac{1}{4}, \\ f(x=0, y=1) = \tfrac{1}{4}, \\ f(x=1, y=0) = \tfrac{1}{4}, \\ f(x=1, y=1) = \tfrac{1}{4}, \end{cases} \tag{3.2}$$

or

$$f(x, y) = \tfrac{1}{4}, \qquad x, y = 0, 1.$$

The basic probability relationships which exist between the single variate frequency functions and the joint function may be best summarized in a table, as in Table 3-1, where the cell entries are the joint frequencies, $f(x, y)$.

Table 3-1. Joint frequency table for tossing two coins

x \ y	0	1	
0	$f(0, 0) = \frac{1}{4}$	$f(0, 1) = \frac{1}{4}$	$f(x=0) = \frac{1}{2}$
1	$f(1, 0) = \frac{1}{4}$	$f(1, 1) = \frac{1}{4}$	$f(x=1) = \frac{1}{2}$
	$f(y=0) = \frac{1}{2}$	$f(y=1) = \frac{1}{2}$	

The main cell entries, then, form a set of mutually exclusive possible occurrences. Adding frequencies along any row gives a probability sum for x fixed and y moving through its possible values. For instance, for the $x = 0$ row, the sum of $f(x = 0, y)$ over columns (y varying) for this example is $f(x = 0, y = 0) + f(x = 0, y = 1)$, that is, the total probability that x takes on the value zero. In more familiar terms, the intersections of $x = 0$ with $y = 0$ and $y = 1$ form two mutually exclusive events whose total probability must be the probability that $x = 0$ no matter what the value of y; i.e., the sum so described represents the unconditional probability that $x = 0$. Analogous interpretations hold for the other row and column sums, as unconditional probabilities.

For this table we have four such relations:

$$\begin{aligned} f(x = 0, y = 0) + f(x = 0, y = 1) &= f(x = 0), \\ f(x = 1, y = 0) + f(x = 1, y = 1) &= f(x = 1), \\ f(x = 1, y = 0) + f(x = 0, y = 0) &= f(y = 0), \\ f(x = 0, y = 1) + f(x = 1, y = 1) &= f(y = 1), \end{aligned} \tag{3.3}$$

which are given as the marginal entries in Table 3-1. Generally speaking, what has been accomplished may be summarized in frequency-function notation as

$$\begin{aligned} f(x) &= \sum_{\text{all } y} f(x, y), \\ f(y) &= \sum_{\text{all } x} f(x, y), \end{aligned} \tag{3.4}$$

for this discrete example. The companion expressions for continuous random variables follow directly:

$$\begin{aligned} f(x) &= \int f(x, y)\, dy, \\ f(y) &= \int f(x, y)\, dx. \end{aligned} \tag{3.5}$$

In both cases they may be called either unconditional frequency functions or *marginal* frequency functions, borrowing from how they are represented in a table of joint frequencies as marginal entries.

The construction of the marginal functions as partial sums (integrals) of joint functions is exactly the same procedure used to form the denominator in Bayes' theorem. And, to complete the analogy, conditional frequency functions may be obtained by the relation

$$f(x \mid y) = \frac{f(x, y)}{f(y)},$$

or (3.6)

$$f(y \mid x) = \frac{f(x, y)}{f(x)}.$$

From Table 3-1, the conditional probabilities are given as the ratios of main cell entries to margins. For $f(y = 0 \mid x = 0)$ we merely divide the (0, 0) joint entry by the $x = 0$ row margin; i.e., $f(y = 0 \mid x = 0) = f(x = 0, y = 0)/f(x = 0)$. Other conditional probabilities are similarly formed.

Statistical independence has a representation in a table of joint probabilities also. The definition of independence of random variables takes on one further refinement, however, in that although joint random variables may possess certain combinations of values for which $f(x, y) = f(x)f(y)$, for the variables to be wholly independent the definition must hold over all such values. For Table 3-1 the variables can easily be seen to be independent, since the joint entries are all products of their respective margins.

As an illustration of nonindependence, consider the drawing of two balls sequentially from an urn containing a mixture of four red balls and one white ball. At each draw the balls remaining in the urn are given equal probabilities of selection. Let x be the random variable for the first draw, with 0, 1 values, corresponding to the selection of a red or white ball, respectively. Similarly, let y represent the same occurrences for the second draw. The joint frequency table and its construction is shown in Table 3-2. Clearly x and y are not independent variables, from, e.g., $f(x = 1, y = 1) \neq f(x = 1)f(y = 1)$.

Table 3-2. Joint frequency table for the urn problem.

x \ y	0	1	
0	$\frac{3}{5}$	$\frac{1}{5}$	$\frac{4}{5}$
1	$\frac{1}{5}$	0	$\frac{1}{5}$
	$\frac{4}{5}$	$\frac{1}{5}$	

$$f(x = 0, y = 1) = f(y = 1 \mid x = 0)f(x = 0) = (\tfrac{1}{4})(\tfrac{4}{5}) = \tfrac{1}{5}$$

$$f(x = 0, y = 0) = f(y = 0 \mid x = 0)f(x = 0) = (\tfrac{3}{4})(\tfrac{4}{5}) = \tfrac{3}{5}$$

$$f(x = 1, y = 1) = f(y = 1 \mid x = 1)f(x = 1) = (0)(\tfrac{1}{5}) = 0$$

$$f(x = 1, y = 0) = f(y = 0 \mid x = 1)f(x = 1) = (1)(\tfrac{1}{5}) = \tfrac{1}{5}$$

The relationships just presented have important uses in a number of applications, among them in the formation of a joint frequency function from the marginal distributions of independent variables, in the derivation of marginal and conditional distributions for situations where the distributions involved are given as explicit mathematical functions, and in deriving one of the three basic distributions from the other two when a convenient joint frequency table cannot be drawn, conceptually—as in the case of continuous random variables—or practically.

To elucidate these suggested uses, consider the joint frequency function $f(x, y) = 2e^{-(2y+x)}$, for $x, y \geq 0$. What are the corresponding marginal distributions? By (3.5),

$$\begin{aligned} f(x) &= \int_0^\infty 2e^{-(2y+x)}\,dy = -e^{-2y-x}\Big|_0^\infty = e^{-x}, \\ f(y) &= \int_0^\infty 2e^{-(2y+x)}\,dx = -2e^{-(2y+x)}\Big|_0^\infty = 2e^{-2y}. \end{aligned} \tag{3.7}$$

Since in this case $f(x)f(y) = f(x, y)$, we deduce that x and y are independent variables.

4. Illustration 1. Discrete probability distributions—the binomial

In practice many probabilistic phenomena may be classified on a priori grounds as following particular theoretical probability distributions. Such is the case when a phenomenon may be described in terms of a finite number of independent, "zero-one" or "failure-success" trials, as in the case of two tosses of a coin. The probability distribution which governs the total number of "failures" or, alternatively, "successes" in the given number of trials is the *binomial distribution*. We have made use of two examples of this distribution previously, in the coin-toss experiment and in our first confrontation with probability, the game of Chapter 1 involving the event "at least one six in four tosses of a die." The decision-rule probability calculations based on two sample games in which the event was either observed or not were binomial probabilities.

To construct the distribution is not difficult. We begin by specifying a random experiment which consists of n independent, identical, success-failure trials, "identical" being used to denote that the probabilities of success and failure are constant over the trials. Another name for such trials is "Bernoulli" trials, after James Bernoulli, a seventeenth-century member of a famous family of mathematicians.

Let us denote the two possible occurrences on each trial via a random variable x_i, $i = 1, \ldots, n$, with probability distribution

$$f(x_i) = f(x) = \begin{cases} p & \text{if } x = 1, \\ 1 - p & \text{if } x = 0. \end{cases} \tag{4.1}$$

The notation $f(x_i) = f(x)$ indicates that the variables x_i possess identical probability distributions. The binomial distribution is concerned with a new random variable, y, the total number of "successes" in n Bernoulli trials.

Since each trial produces $x_i = 1$ in indicating "success," this variable is given by the sum

$$y = \sum_{i=1}^{n} x_i, \tag{4.2}$$

which "counts" the number of successes in n trials.

For example, let us return to the decision-making experiment of Chapter 1. The number of trials, n, is the number of sample observations and we may define "success" as the observation of the event in a play of the game. Then y counts the number of sample games which resulted in occurrence of the event "at least one six in four throws." The main prerequisite to deriving the binomial distribution is to ensure that the probability of "success" remains constant over the trials; i.e., the trials and their corresponding x_i's are independent.

We proceed to construct the distribution by induction. First, what is the probability of no successes in the n trials? This probability is a result of applying the definition of independence for random variables, so that

$$\begin{aligned} P(y=0) &= P(x_1=0)P(x_2=0)\cdots P(x_n=0) \\ &= (1-p)^n = q^n, \end{aligned} \tag{4.3}$$

where $q = 1 - p$, the probability of "failure." To obtain the next term in the distribution, $P(y = 1)$, suppose we look at a simple case of three Bernoulli trials. There are three mutually exclusive ways of getting $y = 1$ for $n = 3$, depicted in Table 3-3. For the first sequence we have $p(q)(q) = pq^2$ as the

Table 3-3. Possible sequences yielding $y = 1$ for $n = 3$

Trial		
1 x_1	**2** x_2	**3** x_3
1	0	0
0	1	0
0	0	1

probability of its occurrence. In fact, this is the probability attached to the individual occurrences of the remaining two sequences. Thus for the $n = 3$ case we obtain

$$P(y = 1 \mid n = 3, p) = 3pq^2. \tag{4.4}$$

We may easily generalize this result by noting that the number of possible sequences involving $(n - 1)$ zeros and exactly (1) one is precisely the combinatorial $\binom{n}{1}$, so that in general

$$P(y = 1 \mid n, p) = \binom{n}{1} pq^{(n-1)}. \tag{4.5}$$

The notation $P(y = k \mid n, p)$ simply refers to the fact that the actual computation of $P(y = k)$ depends on the numerical specification of two constants, the *parameters* n and p.

Further induction shows how the remaining terms of the distribution are developed. For the $y = 2$ case, the probability of any $y = 2$ sequence is clearly p^2q^{n-2}, and since there are exactly $\binom{n}{2}$ ways of producing such sequences,

$$P(y = 2 \mid n, p) = \binom{n}{2} p^2q^{(n-2)}, \tag{4.6}$$

because the sequences are mutually exclusive.

The general term of the binomial distribution is available as

$$P(y = k \mid n, p) = \binom{n}{k} p^kq^{n-k} \tag{4.7}$$

for $y = 0, \ldots, n$, and zero elsewhere. The name of the distribution comes from the binomial expansion of algebra, which is written

$$(a + b)^m = \sum_{r=0}^{m} \binom{m}{r} a^rb^{m-r}. \tag{4.8}$$

The binomial probability distribution contains all the terms in the sum (4.8) for arguments p and q. Thus

$$\sum_{k=0}^{n} \binom{n}{k} p^kq^{n-k} = (p + q)^n = (p + 1 - p)^n = 1, \tag{4.9}$$

which demonstrates that the binomial distribution (4.7) is a true probability distribution.

For a given value of n, the binomial distribution may exhibit varied graphic forms, depending on the value of p. In the coin-tossing example for arbitrary n, $p = \frac{1}{2}$, which generates a perfectly symmetric distribution. Such a result may be easily anticipated, since $p = \frac{1}{2}$ makes *every* term in the distribution possess the common factor $(\frac{1}{2})^n$, so that individual probabilities differ only as the binomial coefficients $\binom{n}{k}$ differ. The coefficients exhibit symmetry since $\binom{n}{k} = \binom{n}{n-k}$, which implies the result.

Figure 3-3 shows three binomial distributions, for $n = 5$, $p = .2, .5, .9$

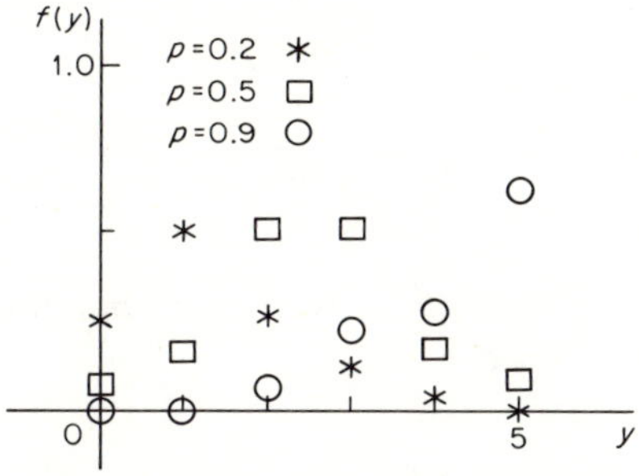

Figure 3-3. The binomial probability distribution $n = 5$.

taken from Appendix Table 1. The binomial tables are given only for $p \leq .5$, since a further symmetry relation shows that $P(y = k \mid n, p) = P(y = n - k \mid n, 1 - p)$. Thus the value for $y = 3$, $p = .9$, $n = 5$ is the same as that for $y = 2$, $p = .1$, $n = 5$, etc.†

One important feature of the binomial distribution is that it exhibits increasing symmetry for any value of p as n increases. This happens to have much practical significance, as it allows the binomial to be approximated by other distributions for large n (cf. the sections below and Chapter 6).

5. Illustration 2. Discrete probability distributions—the Poission

One useful approximation to the binomial distribution under certain conditions and an important distribution in its own right is the Poisson distribution, named after the nineteenth-century French mathematician. As an approximator to the binomial, the Poisson distribution requires p very small, n very large, but such that the product np is a finite constant λ. Specifically, for a series of n Bernoulli trials with probability of success p, the Poisson distribution results when

(a) $p \to 0$,

(b) $n \to \infty$,

(c) such that $np \to \lambda$.

† For the cumulative binomial, the symmetry proposition suggests: $P(y \leq k \mid n, p) = P(y \geq n - k \mid n, q)$, an identity useful in producing brief tables.

Therefore, we are considering "rare" occurrences viewed from the standpoint of an individual trial.

Actual development of the terms in the Poisson distribution may proceed recursively. First, consider the ratio of two adjacent terms in the binomial distribution, say for $y = k$ and $y = k - 1$. The ratio of these probabilities is

$$\frac{P(y = k \mid n, p)}{P(y = k - 1 \mid n, p)} = \frac{\binom{n}{k} p^k q^{n-k}}{\binom{n}{k-1} p^{k-1} q^{n-k+1}} = \frac{np - kp + p}{kq}, \tag{5.1}$$

which simply provides a method by which one binomial probability may be obtained from another, since

$$P(y = k \mid n, p) = \frac{np - kp + p}{kq} P(y = k - 1 \mid n, p). \tag{5.2}$$

For example, using (5.2) on $P(y = 0 \mid n = 3, p = .5) = (\frac{1}{2})^3 = \frac{1}{8}$, we obtain

$$P(y = 1 \mid n = 3, p = .5) = \frac{(3)(.5) - (1)(.5) + (.5)}{(1)(.5)} \times \frac{1}{8} = \frac{3}{8}. \tag{5.3}$$

The analogous "recursion" relation for the Poisson by definition is (5.2) under the conditions (a) to (c). Thus, for $p \to 0 (q \to 1)$, $n \to \infty$, $np \to \lambda$,

$$P(y = k \mid \lambda) = \frac{\lambda}{k} P(y = k - 1 \mid \lambda). \tag{5.4}$$

If we consider the binomial probability for $y = 0$ under these same conditions, then applying (5.4) successively will yield all subsequent terms of the Poisson.

The binomial formula with $y = 0$ gives

$$P(y = 0 \mid n, p) = q^n = \left(1 - \frac{\lambda}{n}\right)^n, \tag{5.5}$$

upon substituting $q = 1 - p$ and $p = \lambda/n$ for the Poisson case. We need to examine (5.5) when $p \to 0$, $n \to \infty$, $np \to \lambda$. Using the Taylor series expansion of $\ln(1 + t)$ for $|t| < 1$ allows an easy evaluation. The expansion is

$$\ln(1 + t) = t - \frac{t^2}{2} + \frac{t^3}{3} - \cdots$$

If we let $t = -\lambda/n$, then the steps involved are as follows:

$$\begin{aligned}\ln P(y=0\mid n,p) &= n\cdot\ln\left(1-\frac{\lambda}{n}\right)\\ &= n\left(-\frac{\lambda}{n}+\frac{\lambda^2}{2n^2}-\frac{\lambda^3}{3n^3}+\cdots\right)\\ &= -\lambda+\frac{\lambda^2}{2n}-\frac{\lambda^3}{3n^2}+\cdots \end{aligned}\tag{5.6}$$

and

$$\lim_{n\to\infty}\ln P(y=0\mid n,p) = -\lambda, \tag{5.7}$$

so that

$$\lim_{n\to\infty} P(y=0\mid n,p) = P(y=0\mid\lambda) = e^{-\lambda}.$$

The bulk of terms in the distribution may now be obtained by applying (5.4) to (5.7); i.e.,

$$\begin{aligned}P(y=0\mid\lambda) &= e^{-\lambda}\\ P(y=1\mid\lambda) &= \frac{\lambda}{1}e^{-\lambda} = \lambda e^{-\lambda}\\ P(y=2\mid\lambda) &= \frac{\lambda}{2}P(y=1\mid\lambda)\\ &= \frac{\lambda^2}{2}e^{-\lambda}\\ P(y=3\mid\lambda) &= \frac{\lambda^3}{(3)(2)}e^{-\lambda}\\ P(y=k\mid\lambda) &= \frac{\lambda^k}{k!}e^{-\lambda} \qquad y=0,1,\ldots\end{aligned}\tag{5.8}$$

as the general term.

The parameter λ being alone in the determination of the above probabilities, as an approximation to the binomial one Poisson distribution may serve many binomial distributions. Thus a Poisson distribution with $\lambda = 3$ approximates all the following distributions, $B(n, p)$†: $B(6, .5)$, $B(15, .2)$, $B(300, .01)$, and many more. According to its underlying assumptions, we

† This notation will be used throughout the text when speaking of a random variable possessing a particular distribution, the exact specification of which depends on certain parameters. Thus $y \sim B(n, p)$ is shorthand for "y is distributed binomially with parameters n and p".

would expect good approximation in general for λ small, otherwise, for large n and small p, as in the $B(300, .01)$ case. These conjectures are borne out by the data of Table 3-4, which shows this Poisson distribution ($\lambda = 3$) versus several of the aforementioned binomials.

Table 3-4. Binomial distributions compared to their Poisson approximation

k	B(6, .5)	B(30, .1)	B(150, .02)	B(300, .01)	Poisson $\lambda = 3$
0	.0156	.0424	.0483	.0490	.0498
1	.0937	.1413	.1478	.1486	.1494
2	.2344	.2276	.2248	.2244	.2240
3	.3125	.2361	.2263	.2252	.2240
4	.2344	.1771	.1697	.1689	.1680
5	.0937	.1023	.1011	.1010	.1008
6	.0156	.0474	.0499	.0501	.0504
7		.0180	.0209	.0213	.0216
8		.0058	.0076	.0079	.0081
9		.0016	.0025	.0026	.0027
10		.0004	.0007	.0008	.0008
11		.0001	.0002	.0002	.0002
12		.0000	.0000	.0000	.0001

The Poisson has a definite and important meaning independent of its use as an approximator of the binomial distribution, as the probability law for the occurrence of stochastic (probabilistic) events over time or space. Many physical phenomena, such as ray emissions, may be viewed as being governed by a Poisson probability law. The main assumption required to produce the Poisson distribution in such cases is that the "process" be time-independent; i.e., if you observe a shooting star in some fixed area of space at one instant, the probability that another is sighted in the next instant of time is independent of your first observation. For any span of time, then (for convenience say an interval of unit length), the interval is divided up into n subintervals of length $1/n$. It is assumed that each subinterval is such as to allow at most one occurrence—that is, the probabilities of more than one occurrence are assumed zero. Clearly such an assumption will be met in the limit, as $n \to \infty$. Since the probabilities in each subinterval are presumed constant over the subintervals we have defined precisely the situation under which the binomial becomes the Poisson.

Extension to an arbitrary time interval, t, follows directly by dividing the interval into n subintervals and proceeding as above. Then in equations (5.4) and (5.5) λ/k and λ/n become $(t\lambda)/k$ and $(t\lambda)/n$. The time-oriented Poisson follows as

$$P(y = k \mid \lambda t) = \frac{(\lambda t)^k}{k!} e^{-\lambda t}. \tag{5.9}$$

Applications of this generalized Poisson law are many and varied. It has been used to describe, e.g., the behavior of incoming calls to a telephone switchboard, and has important use in the analysis of waiting-line problems.† Its closeness of approximation to phenomena which, on face, cannot be treated as stochastic in nature but nonetheless exhibit Poisson behavior is sometimes startling.‡

6. Illustration 3. Continuous probability distributions—the normal

The binomial distribution may be approximated with a continuous probability distribution, as may the Poisson. The famous Central Limit Theorem of statistics states that for a set of n independent random variables $[x_1, \ldots, x_n]$, no matter what their distribution, the sum $y = \sum_{i=1}^{n} x_i$ has the normal distribution in the limit as $n \to \infty$. Of course the binomial variable y fits this situation precisely, so that for large n we should expect that the normal is a good approximation to the binomial.

The normal distribution has as its density function the form§

$$f(x) = \frac{1}{\sigma\sqrt{2\pi}} \exp\left[\frac{-\frac{1}{2}(x-\mu)^2}{\sigma^2}\right], \tag{6.1}$$

which produces the familiar bell-shaped curve of Figure 3-4, where the two parameters of the distribution are set arbitrarily at $\mu = 5$, $\sigma = 1$. Using the notation introduced previously, we would speak of the variable x for this distribution as $x \sim N(\mu, \sigma)$.

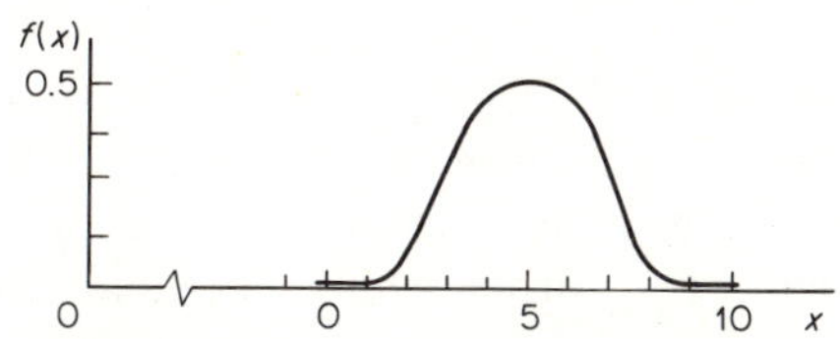

Figure 3-4. The normal distribution, $N(5, 1)$.

As an approximator of discrete distributions the normal cannot, being continuous, provide probability values for particular values of x (cf. Section 1, this chapter). It can, however, approximate the distribution function of, e.g.,

† See, e.g., P. M. Morse, *Queues, Inventories, and Maintenance* (Wiley, 1958).

‡ One very famous example is in the description of the outbreak of wars during certain periods of time. See L. F. Richardson, "The Statistics of Deadly Quarrels," in J. R. Newman (ed.), *The World of Mathematics* (Simon & Schuster, 1956), Vol. II, pp. 1254-1263.

§ The term "exp" means "e raised to the power."

the binomial variable y. Such a comparison is shown in Table 3-5 for the normal approximation to the $B(300, .01)$ distribution. Exactly how such an approximation is obtained and, indeed, how one uses the normal tables (Appendix Table 2) will be discussed presently. The particular normal distribution to use as the approximation in any given case depends on a translation of the binomial parameters n and p into their normal counterparts μ and σ. And its ability to approximate closely depends on when n is "large enough." Table 3-5 suggests that for values of p close to zero or 1 in

Table 3-5. The normal approximation to $B(300, .01)$; Cumulative probability $P(y \leq k)$

k	B(300, .01)	N(3, 1.64)
0	.0490	.0336
1	.1976	.1112
2	.4220	.2209
3	.6472	.5000
4	.8161	.7291
5	.9171	.8888
6	.9672	.9664
7	.9885	.9927
8	.9964	.9989
9	.9990	1.0000
10	.9998	1.0000
11	1.0000	1.0000

the binomial, n must be very large before the normal approximation is accurate. When the binomial is close to being symmetric ($p \simeq \frac{1}{2}$), even relatively small values for n allow the normal to provide a satisfactory approximation to the binomial. The former statement holds in the Poisson case: n must be very large.

It is difficult at this point in the development to *argue* the significance of the normal distribution in statistics, except from the limit theorem given above. Such will become apparent as we go on. The distribution has some significance in its own right, being the empirical distribution of certain attributes in human populations, such as weight, height, etc. Other examples exist, and will be alluded to as appropriate in the material to come.

4

Moments of single variate distributions

The binomial distribution rests on certain theoretical assumptions, as does the Poisson, but no such convenient check exists for the normal distribution. For instance, given independent Bernoulli trials, the random variable y, number of successes in n trials, must follow the binomial distribution. Other than being a limiting distribution for many probability distributions, the normal distribution can apply only by assumption, the validity of the assumption to be checked by confrontation with the real-world phenomenon it purports to describe.

Indeed, a major problem for applying the subject matter to follow exists in how to identify (given that a theoretical distribution cannot be inferred) the underlying probability distribution. This is not to say that the major problem of inference is concerned with the identification of the underlying probability distribution. Most statistical inference, in fact, is of more limited scope. It is concerned with inferences about parameters *in some known distribution*, as if that problem (identity of the underlying distribution) had been solved previously. Ignorance of this basic fact results in many instances of faulty application, which result, in turn, in bad inference.

How does one identify a particular distribution from actual data? And, how can one ever be sure of this identification? The answer to the latter question is simply, "one can never be sure," unless, as mentioned before, there are indications that a particular theoretical distribution holds. The uncertainty derives from (1) the limitations of our techniques for identification and (2) the reality that most information about any given phenomenon is in the form of *sample* information. Thus it is in general true that we can never identify a particular distribution from empirical data with certainty.

All frequency distributions have, besides their parameters or arguments, other characteristic values, called *moments*. For most cases these moments are finite quantities and give, in descriptive terms, indications as to how the random variable or functions of it are distributed around some constant in the range of the random variable. Thus a given distribution will possess certain unique moments which may be compared to their surrogates from actual observation. The ability to identify a particular distribution from such actual data is therefore limited only by the uniqueness of a set of moments and our ability to compare them with actual observation.

Other uses for certain moments in frequency distributions exist and are important for our future work in statistical inference. They will be introduced as we proceed.

1. Expected value—the arithmetic mean

All moments in a frequency distribution are computed as functions of the probabilities of the random variable involved. The basic definition underlying their computation is that of *mathematical expectation* or *expected value*. If $g(x)$ is some rational function of the random variable x, and $f(x)$ is the probability distribution of x, then the general version of the "expected value of $g(x)$," denoted $E[g(x)]$, is given by

$$E[g(x)] = \begin{cases} \int g(x)f(x)\,dx\text{: continuous,} \\ \sum_{\text{all } x} g(x)f(x)\text{: discrete.} \end{cases} \tag{1.1}$$

The most familiar expected value is $E(x)$, i.e., where $g(x) = x$, the "expected value of x" or its arithmetic mean, μ. The definition (1.1) gives the formula for μ as

$$\mu = E(x) = \begin{cases} \int xf(x)\,dx: \text{ continuous}, \\ \sum_{\text{all } x} xf(x): \text{ discrete}. \end{cases} \tag{1.2}$$

Here μ is also the first moment of x around zero. The general form for all moment calculations is (1.1) with $g(x - k)$ inserted, where k is some arbitrary constant.

It should be obvious that μ will lie in the range of x, but it is not at all obvious that μ must also be an attainable or possible value of x. For the continuous case this must be so, but in the discrete case it need not be.

For purposes of illustration, let us compute $E(x)$ for a single Bernoulli trial using (1.2). If the probability of success is p, then (1.2) yields

$$\begin{aligned} \mu = E(x) &= 0 \cdot P(x = 0) + 1 \cdot P(x = 1) \\ &= 0 \cdot q + 1 \cdot p \\ &= p \end{aligned} \tag{1.3}$$

If a simple coin toss is the Bernoulli random variable, (1.3) says that the "average" value of x is $\frac{1}{2}$. Of course, any given trial can result in either $x = 0$ or $x = 1$ but never $x = \frac{1}{2}$. The interpretation must be in terms of a long-run connotation, with a meaning similar to the concept of probability. For instance, were you to bet \$1 on the event "heads" for many, many games of tossing a coin once, you would "expect" to win half the time and lose \$1 the other half, and so break even. The same analysis applies to the game of Chapter 1, where you would expect not to break even but to win about 4¢ per dollar bet, given a fair die and betting always on the event's occurrence.

Descriptively speaking, μ is a "central" value of $f(x)$, or μ measures *central tendency* in the distribution of x, not unlike the establishment of the fulcrum point for a series of "mass points" on the real line. Thus the "effective" weight of a mass in relation to the balance point is a combination (product) of its weight and distance from the fulcrum point. The point which "balances" the line (equalizes the downward forces) is μ.

The expected value of a random variable owns an important place in this book, because the principles of statistical inference are presented almost entirely from the standpoint of inferences to a mean μ. To be able to calculate expected values for functions of random variables (such as the binomial variable, a sum of n other random variables) is required, and rests on the development of such calculation rules for expected values as we presented previously for probabilities. We shall proceed to give these rules of calculation for expected values via the binomial distribution.

What is $\mu = E(y)$ for the binomial? By definition (1.2), μ may be calculated as

$$\mu = E(y) = \sum_{y=0}^{n} y \binom{n}{y} p^y q^{n-y}. \tag{1.4}$$

The steps involved in evaluating (1.4) may be outlined as follows:

$$E(y) = \underbrace{0 \cdot \binom{n}{0} p^o q^n}_{\text{1st term}=0} + \sum_{y=1}^{n} y \binom{n}{y} p^y q^{n-y}. \tag{1.5}$$

Canceling y into the combinatorial,

$$E(y) = \sum_{y=1}^{n} \frac{n!}{(n-y)!(y-1)!} p^y q^{n-y}. \tag{1.6}$$

Factor np out of the sum:

$$E(y) = np \sum_{y=1}^{n} \frac{(n-1)!}{(n-y)!(y-1)!} p^{y-1} q^{n-y}. \tag{1.7}$$

Rewrite with $z = y - 1$, $m = n - 1$:

$$E(y) = np \underbrace{\sum_{z=0}^{m} \binom{m}{z} p^z q^{m-z}} = np. \tag{1.8}$$

The underscored portion of (1.8) is the binomial expansion of $(p+q)^m = 1$, thus $E(y) = np$; the average value of the number of successes in n trials equals the number of trials times the probability of success (also the expected value) on each. This result suggests that for random variables which are constructed as sums of other random variables, the expectation of the sum variable equals the sum of expectations; i.e.,

Rule 1. *Expected value of a sum of random variables.* If x_i, $i = 1, \ldots, n$ are random variables, with $y = \sum_{i=1}^{n} x_i$, then

$$E(y) = \sum_{i=1}^{n} E(x_i). \tag{1.9}$$

We shall prove rule 1 by induction from the result for a particular sum of random variables to the general case, via a careful analysis of the relation of expectation in frequency function terms to expectation expressed in terms of the original sample-space probabilities.

Consider a (finite) sample space for the variable y in $y = x + z$, defined by the "points" u_i, $i = 1, \ldots, n$; i.e., $S = \{u_i\}$. The probabilities associated with these elementary events are $\{p_i\}$, and we may write the random variables y, x, and z as functionally related to the $\{u_i\}$ as $y(u_i) = (x + z)(u_i)$, $x(u_i)$, and $z(u_i)$. It will be shown first that the definition of $E(y)$ [i.e., equation (1.2)] may easily be expressed in terms of the sample space as

$$E(y) = E(x + z) = \sum_{i=1}^{n} (x + z)(u_i)p_i . \tag{1.10}$$

That (1.10) is an equivalent way of writing the expected value of y is clear from the definition of the frequency function. But to emphasize this correspondence we shall examine the relationship in detail via an example. This presentation will also produce the key to proving rule 1.

For the binomial case, consider a set of three Bernoulli trials, where a coin is tossed. Schematically we may represent the sample space as a set of eight elementary events, intersections of the two-value random variables on each Bernoulli trial, as in Table 4-1. If the sample-space probabilities are,

Table 4-1. Sample space and binomial random variable for $n = 3$

	Bernoulli random variable						
$y = \sum_{1}^{3} x_i$	x_1		x_2		x_3		
$y = 0$	0	$\cap$	0	$\cap$	0	U_1	elementary events S
	1	$\cap$	0	$\cap$	0	U_2	
$y = 1$	0	$\cap$	1	$\cap$	0	U_3	
	0	$\cap$	0	$\cap$	1	U_4	
	1	$\cap$	1	$\cap$	0	U_5	
$y = 2$	1	$\cap$	0	$\cap$	1	U_6	
	0	$\cap$	1	$\cap$	1	U_7	
$y = 3$	1	$\cap$	1	$\cap$	1	U_8	

correspondingly, $p_1, \ldots, p_8$, then the frequency distribution of y, $f(y)$, is as follows: $f(y = 0) = p_1$, $f(y = 1) = p_2 + p_3 + p_4$, $f(y = 2) = p_5 + p_6 + p_7$, and $f(y = 3) = p_8$. From the frequency function definition of expectation,

$$E(y) = \sum_{y=0}^{3} yf(y) = 0 \cdot f(y = 0) + 1 \cdot f(y = 1) + 2f(y = 2) + 3f(y = 3). \tag{1.11}$$

Consider any one of the terms in (1.11), say $1 \cdot f(y = 1)$. In terms of the sample space $1f(y = 1) = 1(p_2 + p_3 + p_4) = 1p_2 + 1p_3 + 1p_4$. A similar representation may be written for the remaining terms, so that clearly

$$E(y) = \sum_{y=0}^{3} yf(y) = \sum_{i=1}^{8} y(u_i)p_i. \tag{1.12}$$

Each $y(u_i)$ can be decomposed into constituent parts, since, e.g.,

$$y(u_2) = 1 = \sum_{j=1}^{3} x_j(u_2).$$

That is, each x has a particular representation for each u_i, as well indeed it must. Inserting this relationship in (1.12), we obtain

$$\begin{aligned} E(y) &= \sum_{i=1}^{8} y(u_i)p_i \\ &= \sum_{i=1}^{8} \sum_{j=1}^{3} x_j(u_i)p_i \\ &= \sum_{j=1}^{3} \left[\sum_{i=1}^{8} x_j(u_i)p_i \right] \\ &= \sum_{j=1}^{3} [E(x_j)], \end{aligned} \tag{1.13}$$

by definition. The result may easily be generalized by induction, inserting for any x_j a sum of other random variables, and holds for continuous random variables as well as discrete ones.

Using rule 1 on the binomial problem,

$$\mu = E(y) = \sum_{i=1}^{n} E(x_i) = \sum_{i=1}^{n} p = np. \tag{1.14}$$

From the specifications underlying the Poisson distribution it should be clear that since $np = \lambda$, the mean of the Poisson is λ. This can easily be shown by applying the methods of (1.4) to (1.8) to the Poisson.

Other useful rules for calculations involving expected values are as follows:

Rule 2. *Expected value of a constant times a random variable.* If x is a random variable and $y = kx$ for k a scalar,

$$E(y) = kE(x). \tag{1.15}$$

Rule 2 is easy to demonstrate by writing the expectation in either of its definitional forms and factoring out k. This process also shows that $E(k) = k$.

Rule 3. *Expected value of a product of random variables.* If x and z are *independent* random variables, then

$$E(x \cdot z) = E(x)E(z). \tag{1.16}$$

To show rule 3 let us consider the definition of $E(x \cdot z)$ in terms of the joint frequency function $f(x, z)$. Then

$$E(x \cdot z) = \sum_{\text{all } x,\, z} (x \cdot z) f(x, z)$$

for $x \cdot z$ discrete. It follows that the sum is, in actuality, a double sum, so that

$$E(x \cdot z) = \sum_{\text{all } x} \sum_{\text{all } z} (x \cdot z) f(x, z). \tag{1.17}$$

If x and z are independent, $f(x, z) = f(x)f(z)$ and

$$E(x \cdot z) = \sum_{\text{all } x} \sum_{\text{all } z} (x \cdot z) f(x) f(z). \tag{1.18}$$

Multiplying and sorting,

$$\begin{aligned} E(x \cdot z) &= \left[\sum_{\text{all } x} x f(x)\right]\left[\sum_{\text{all } z} z f(z)\right] \\ &= E(x)E(z), \end{aligned} \tag{1.19}$$

as required. In general, therefore, $E(x \cdot z) \neq E(x)E(z)$, with equality only under conditions of statistical independence.

As an application of rule 3, consider $z = x \cdot y$, where x and y are independent identically distributed Bernoulli variables, corresponding to the "product" results of a coin toss, where $z = 1$ only if both trials result in "success" and $z = 0$ otherwise. If p is the probability of "success" on each trial, then the probability distribution of z is:

(a) $P(z = 0) = P(y = 0, x = 0) + P(y = 1, x = 0) + P(y = 0, x = 1)$
$= q^2 + 2pq;$

(b) $P(z = 1) = P(y = 1, x = 1) = p^2.$

The definition of expectation gives

$$\begin{aligned} E(z) &= (0)P(z = 0) + (1)P(z = 1). \\ &= p^2. \end{aligned} \tag{1.20}$$

The same result is available directly from rule 3, since $E(x) = E(y) = p$ and x and y are independent:

$$\begin{aligned} E(z) &= E(x)E(y) \\ &= p^2. \end{aligned} \tag{1.21}$$

2. Moments of distributions—the variance

The arithmetic mean $E(x) = \mu$ does fall within our definition of the *moments* of frequency distributions, i.e., $E[g(x-k)]$, where $k = 0$ and $g(x) = x$. Whenever $k = 0$ we are calculating moments about the origin for the distribution, the first of which is the mean. For a general definition, we denote the kth moment of a distribution around zero as μ_k^0,

$$\mu_k^0 = \begin{cases} \sum_{\text{all } x} x^k f(x)\text{: discrete,} \\ \int x^k f(x)\, dx\text{: continuous.} \end{cases} \tag{2.1}$$

As descriptive devices these moments about zero may not be appropriate for a given distribution, e.g., when the distribution does not have zero as a possible value or has its complete range to one or the other side of zero. Of course, the arithmetic mean is the first-order moment about zero, and has special significance as a descriptive measure. Higher-order moments, however, may not be of such obvious descriptive usefulness. But for purposes of otherwise identifying a distribution, they provide the necessary basis for comparison to empirical observation.

Of more descriptive usefulness are moments which show how the variable is distributed around a central value, such as the mean. These "moments around the mean," denoted μ_k, are defined via our previous specification by

$$\mu_k = \begin{cases} \sum_{\text{all } x} (x-\mu)^k f(x)\text{: discrete,} \\ \int (x-\mu)^k f(x)\, dx\text{: continuous.} \end{cases} \tag{2.2}$$

The second-order moment μ_2 has special significance in statistics, and is called the *variance* of the distribution $f(x)$.† As a descriptive measure, the

† The first-order moment around μ is zero:

$$\begin{aligned} \mu_1 = \sum_{\text{all } x} (x-\mu) f(x) &= \sum_{\text{all } x} x f(x) - \mu \sum_{\text{all } x} f(x) \\ &= \mu - \mu(1) \\ &= 0. \end{aligned}$$

variance yields an indication of how the variable, x, is dispersed about its mean, μ. And for our purposes the variance provides an operational definition of the precision of an estimate. For example, if x is an estimate of μ, based on a random sampling procedure which generates $f(x)$ as the distribution of x, then x is a perfectly precise estimate of μ if $x = \mu$ always. Degrees of imprecision may be defined via some measure of dispersion of x from μ, and a convenient measure for this purpose is the variance, $\mu_2 = \sigma^2$†:

$$\sigma^2 = \begin{cases} \sum_{\text{all } x} (x - \mu)^2 f(x)\text{: discrete,} \\ \\ \int (x - \mu)^2 f(x)\, dx\text{: continuous.} \end{cases} \tag{2.3}$$

As an illustration of the descriptive usefulness of σ^2, consider the ordinary decision problem facing a buyer of automobile tires. For simplicity assume the choice is between two brands at identical cost, where for brand X tire wear is quoted with a mean $\mu_x = 21{,}000$ miles and for brand Y, $\mu_y = 23{,}500$ miles. Clearly brand Y is superior on the basis of its average wear, but our buyer is going to purchase only one set of tires, not many different sets. Thus the average has meaning to him but does not totally describe the situation. By the definition of σ^2, the variance yields an "average" also (as do all moments), the average squared deviation of the actual wear figures from the average, μ. So if brand X had $\sigma_x^2 = 250{,}000$ miles^2 and brand Y had $\sigma_y^2 = 1{,}000{,}000$ miles^2, we would judge brand X superior on this count.

The usual convention is to quote dispersion not in terms of σ^2 but in terms of $\sigma = \sqrt{\sigma^2}$, called the *standard deviation*. For the above figures, $\sigma_x = 500$ miles and $\sigma_y = 1000$ miles. These values have more meaning in relation to their respective means, and might represent the situation depicted in Figure 4-1.

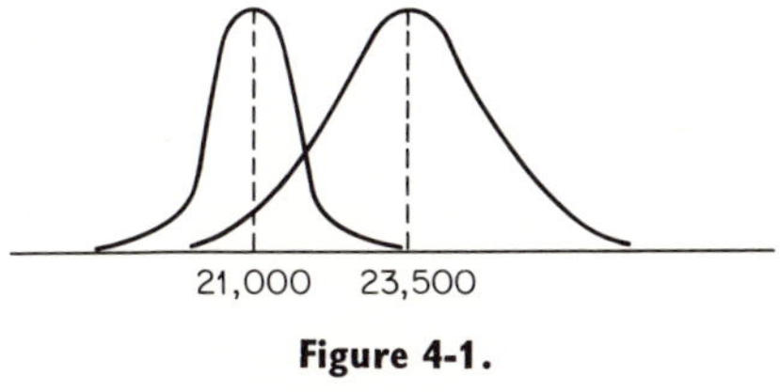

Figure 4-1.

From a purely descriptive point of view, one might suggest the average of absolute deviations, or some other measure, in lieu of σ. Indeed, in this light σ is purely arbitrary. However, in our future work the second moment

† Recall the notation σ^2 as a parameter of the normal distribution (cf. Section 6, Chapter 3). In the normal case σ^2 is the variance of the distribution and coincidentally appears as an explicit parameter in the density function $f(x)$. In general the moments of a distribution are not its arguments.

σ^2 will be of primary importance. So conventionally we use σ as the measure of dispersion in a distribution.

Figure 4-1 is not the only possible representation for the actual wear distributions of brand X and brand Y. If it were, then the customer may choose one brand over the other on the basis of average wear versus deviations of individual tires only. That is not to say such a choice is clearly in favor of one brand or the other, but further confusion might be added if the actual distributions were those of Figure 4-2.

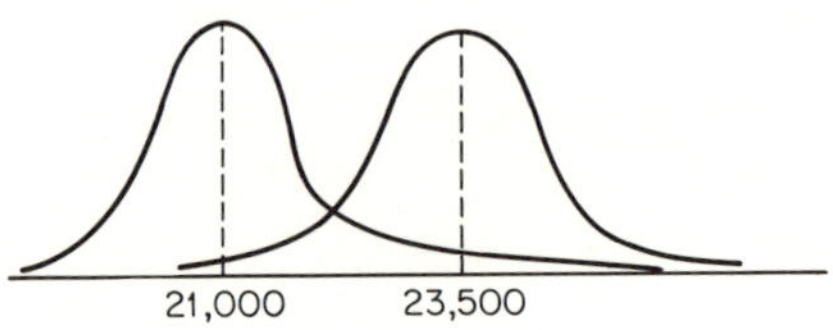

Figure 4-2.

In such a case as this dispersion as measured by σ is not a sufficient indicator of the situation, since σ connotes nothing about the *skewness* or asymmetry of the two distributions. What is required to indicate skewness is a measure which takes into account the signs and algebraic magnitudes of deviations from μ as σ^2 does not. As generally used, the third moment around μ is an indicator of the skewness of a distribution.

For calculation purposes, all moments around the mean may be related to their lower-order counterparts and moments around zero. The variance, for example, may be represented in terms of the second moment around zero and the mean, as in the discrete case:

$$\begin{aligned}\mu_2 = \sigma^2 &= \sum_{\text{all } x} (x - \mu)^2 f(x) \\ &= \sum_{\text{all } x} x^2 f(x) - 2\mu \sum_{\text{all } x} x f(x) + \mu^2 \sum_{\text{all } x} f(x) \\ &= \mu_2^0 - 2\mu(\mu) + \mu^2(1) \\ &= \mu_2^0 - \mu^2. \end{aligned} \tag{2.4}$$

In general all higher-order moments around the mean will be available in terms of the corresponding moment around zero and lower-order moments around the mean.

3. Some theorems on variances

As in the case of mean calculations, when random variables are functions of other random variables we may expedite the calculation of a variance by using some general rules for such calculations.

Consider first the variance of the simple Bernoulli variable. According to our previous work, if x is the Bernoulli variable, then $\mu = E(x) = p$, the probability of success on the trial. Equation (2.3) gives the variance σ^2 in one form, which by virtue of the definition of expectation is equivalent to $E(x-\mu)^2$. This formula may be further expanded to give [as in (2.4)]

$$\sigma^2 = E(x-\mu)^2 = E(x^2) - \mu^2. \tag{3.1}$$

Applying (3.1) to find σ^2 in the case of a single Bernoulli trial,

$$\begin{aligned} E(x^2) &= 1^2 \cdot p + 0^2 \cdot q \\ &= p, \end{aligned} \tag{3.2}$$

so that

$$\sigma_x^2 = p - p^2 = p(1-p) = pq. \tag{3.3}$$

If a coin is fair, therefore, $E(x) = \mu = \frac{1}{2}$, $\sigma_x^2 = \frac{1}{4}$, $\sigma_x = \frac{1}{2}$.

We may use the general expression for the variance to obtain σ_y^2 in the binomial distribution, with the hope that the result obtained may suggest a general rule for calculating the variance of a random variable which is the sum of other random variables. We begin by using (3.1):

$$\sigma_y^2 = E(y^2) - \mu_y^2, \tag{3.4}$$

where $y = \sum_{i=1}^{n} x_i$ and $E(y) = np$ as before. Proceeding in outline fashion,

$$\begin{aligned} \sigma_y^2 &= \sum_{y=0}^{n} y^2 f(y) - (np)^2 \\ &= \sum_{y=0}^{n} y^2 \frac{n!}{(n-y)!\,y!} p^y q^{n-y} - (np)^2. \end{aligned} \tag{3.5}$$

Now substitute $y^2 = y(y-1) + y$, to get

$$\begin{aligned} \sigma_y^2 &= \sum_{y=0}^{n} y(y-1) \frac{n!}{(n-y)!\,y!} p^y q^{n-y} + \sum_{y=0}^{n} y \frac{n!}{(n-y)!\,y!} p^y q^{n-y} - (np)^2 \\ &= \sum_{y=0}^{n} y(y-1) \frac{n!}{(n-y)!\,y!} p^y q^{n-y} + np - (np)^2, \end{aligned} \tag{3.6}$$

since the second term is equal to $E(y) = np$. We note that in the remaining sum the first two terms are zero, so that (3.6) reduces, upon canceling $y(y-1)$ into $y!$ to

$$\sigma_y^2 = \sum_{y=2}^{n} \frac{n!}{(n-y)!\,(y-2)!} p^y q^{n-y} + np - (np)^2. \tag{3.7}$$

Using the trick involved in obtaining $E(y)$ by brute force [cf. equations (1.6) and (1.7)], if we factor $n(n-1)p^2$ from the sum we are left with

$$\sigma_y^2 = n(n-1)p^2 \sum_{y=2}^{n} \frac{(n-2)!}{(n-y)!\,(y-2)!} p^{y-2} q^{n-y} + np - (np)^2. \tag{3.8}$$

Let $m = n - 2$ and $z = y - 2$. Then

$$\begin{aligned}\sigma_y^2 &= n(n-1)p^2 \sum_{z=0}^{m} \frac{m!}{(m-z)!\,z!} p^z q^{m-z} + np - (np)^2 \\ &= n(n-1)p^2(1) + np - (np)^2 \\ &= np(1-p) \\ &= npq. \end{aligned} \tag{3.9}$$

Since each individual trial has a variance $\sigma_x^2 = pq$, then npq represents a sum of these variances over the n trials. Thus a plausible generalization is that the variance of a sum of random variables is the sum of individual variances. As a matter of fact, this generalization does not hold except under the assumption of independence mutually among the summed variables:

Rule 1. *Variance of a sum of random variables.* If $y = \sum_{i=1}^{n} x_i$, with σ_i^2 as the variance of x_i and the $\{x_i\}$ mutually independent, then

$$\sigma_y^2 = \sum_{i=1}^{n} \sigma_i^2. \tag{3.10}$$

The proof of rule 1 and the development of the analogous theorem without the independence assumption follows easily with induction. Consider first the random variable $z = x + y$. Then by definition,

$$\sigma_z^2 = E[z - E(z)]^2 = E[(x+y) - E(x+y)]^2. \tag{3.11}$$

Rule 1 for expectations yields

$$\begin{aligned}\sigma_z^2 &= E[(x+y) - E(x) - E(y)]^2 \\ &= E\{[x - E(x)] + [y - E(y)]\}^2 \\ &= E[x - E(x)]^2 + E[y - E(y)]^2 + 2E[x - E(x)][y - E(y)], \end{aligned} \tag{3.12}$$

by expanding the quadratic and applying rule 1 for expectations once again. Now the first term in the result of (3.12) is σ_x^2, the second is σ_y^2, and the third involves the *covariance* of x and y, which we shall denote σ_{xy}. Thus

$$\sigma_z^2 = \sigma_x^2 + \sigma_y^2 + 2\sigma_{xy}.\dagger \tag{3.13}$$

Expanding the covariance term,

$$\begin{aligned}\sigma_{xy} &= E[x - E(x)][y - E(y)] \\ &= E[xy - xE(y) - yE(x) + E(x)E(y)] \\ &= E(xy) - 2E(x)E(y) + E(x)E(y) \\ &= E(xy) - E(x)E(y).\end{aligned} \tag{3.14}$$

From rule 3 on expectations, $E(xy) = E(x)E(y)$ only if x and y are *independent*. Thus under conditions of independence, $\sigma_{xy} = 0$‡ and

$$\sigma_{x+y}^2 = \sigma_{x-y}^2 = \sigma_x^2 + \sigma_y^2. \tag{3.15}$$

To generalize (3.15) we merely put $x = u + v$, where u and v are two other independent random variables, etc.

The nonindependence case may be extended by a similar substitution, $x = u + v$. Then (3.12) becomes

$$\sigma_{u+v+y}^2 = E[x - E(x)]^2 + \sigma_y^2 + 2E[x - E(x)][y - E(y)]. \tag{3.16}$$

The first term, $E[x - E(x)]^2 = \sigma_u^2 + \sigma_v^2 + 2\sigma_{uv}$, from rule 1 for variances. The covariance term σ_{xy} is expanded as

$$\begin{aligned}\sigma_{xy} &= E\{[u - E(u)][y - E(y)] + [v - E(v)][y - E(y)]\} \\ &= \sigma_{uy} + \sigma_{vy},\end{aligned} \tag{3.17}$$

and (3.16) becomes

$$\sigma_{u+v+y}^2 = \sigma_u^2 + \sigma_v^2 + \sigma_y^2 + 2\sigma_{uv} + 2\sigma_{uy} + 2\sigma_{vy}. \tag{3.18}$$

Further extension shows the general form for a sum of n variables to be

$$\sigma_\Sigma^2 = \sum(\text{individual variances}) + \binom{n}{2}\text{covariances}. \tag{3.19}$$

† Were we to have begun with $z = x - y$ we would find that $\sigma_z^2 = \sigma_x^2 + \sigma_y^2 - 2\sigma_{xy}$.

‡ It is not in general true, however, that $\sigma_{xy} = 0$ implies the independence of x and y. An example illustrating this fact is given in Appendix 1 to this chapter.

Rule 2 for expectations also has a counterpart for variance calculations. Consider $z = ky$, where k is a scalar and y is a random variable with σ_y^2 as its variance. Then

$$\sigma_z^2 = E[z - E(z)]^2 = E[ky - kE(y)]^2, \tag{3.20}$$

using rule 1 for expectations to write $E(ky) = kE(y)$. The constant may be factored from the above expression to yield

$$\begin{aligned}\sigma_z^2 &= k^2 E[y - E(y)]^2 \\ &= k^2 \sigma_y^2,\end{aligned} \tag{3.21}$$

which we state more formally as

Rule 2. *Variance of a product of a scalar and a random variable.* For $z = ky$, with k a scalar and y a random variable with variance σ_y^2,

$$\sigma_z^2 = k^2 \sigma_y^2 .$$

If we combine these two rules of variance calculation, it is clear that for $y = \sum_{i=1}^n w_i x_i$ and the x_i's mutually independent random variables,

$$\sigma_y^2 = \sum_{i=1}^{n} w_i^2 \sigma_i^2 , \tag{3.22}$$

where the individual variances of the x_i's are denoted σ_i^2.

For the binomial random variable all the $w_i = 1$, and the variances are each equal to pq. Thus, since the individual Bernoulli trials are independent, $\sigma_y^2 = npq$. The variance for the Poisson distribution may easily be ascertained by inspection. Since under the conditions which generate the Poisson from a binomial situation we have $np \to \lambda$, and $q \to 1$, clearly for $y \sim P(\lambda)$, $\sigma_y^2 = \lambda$. The Poisson mean and variance are equal.

There is no analogous calculation rule for the variance of a product (quotient) of random variables. The reason this difficulty arises can be seen easily: For $z = x \cdot y$,

$$\begin{aligned}\sigma_{x \cdot y}^2 &= E[xy - E(xy)]^2 \\ &= E[(xy)^2] - [E(xy)]^2.\end{aligned} \tag{3.23}$$

Even with an independence assumption the expansion of $E[(xy)^2]$ is not available. $\sigma_{x \cdot y}^2$ may, however, be approximated to an arbitrary degree of accuracy; the presentation of the approximation is beyond the scope and need of this book, and is given elsewhere.†

† For the case of a product of two random variables, see Hansen, M., Hurwitz, L. and Madow, W., *Sample Survey Methods and Theory*, Vol. II (Wiley, 1953), pp. 107–109.

A note on independence and zero covariance

From equation (3.14), the independence of two variables, x and y, implies that their covariance $\sigma_{xy} = 0$. The reverse implication is not, in general, true, as seen by this counterexample.† Consider a sample space consisting of three points and as given in Figure 4A-1, with a probability structure as shown. If these variables are

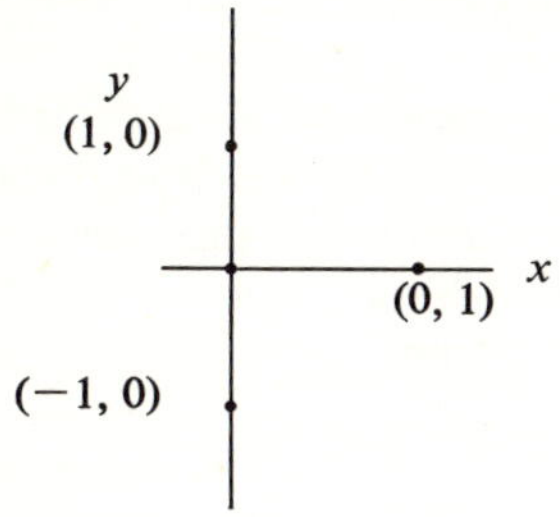

y \ x	0	1	f(y)
−1	$\frac{1}{4}$	0	$\frac{1}{4}$
0	0	$\frac{1}{2}$	$\frac{1}{2}$
1	$\frac{1}{4}$	0	$\frac{1}{4}$
f(x)	$\frac{1}{2}$	$\frac{1}{2}$	1

Figure 4A-1.

independent, then $f(x, y) = f(x)f(y)$ for all x and y. Independence does not hold, since, e.g., $f(x = 0, y = -1) \neq f(x = 0)f(y = -1)$, i.e., $[\frac{1}{4} \neq (\frac{1}{2})(\frac{1}{4})]$.

The covariance σ_{xy} may easily be calculated from the definition (3.14). For this example, $E(x) = (0)(\frac{1}{2}) + (1)(\frac{1}{2}) = \frac{1}{2}$, $E(y) = (-1)(\frac{1}{4}) + (0)(\frac{1}{2}) + (1)(\frac{1}{4}) = 0$, and $E(xy) = 0$, so that $\sigma_{xy} = 0$.

† The author wishes to thank Prof. T. A. Yancey for this example.

Standard normal distribution

To tabulate the great many normal probability distributions according to the possible combinations of parameter values for μ and σ^2 would indeed be an enormous task, besides requiring many bound volumes for purposes of presentation. Fortunately such an epic work is not necessary. Instead, it is possible to obtain probabilities for any normal distribution from any other, through a simple linear transformation of the variable involved. Hence Appendix Table 2 contains values for only one normal distribution.

What normal distribution should be used for purposes of constructing tables? According to these remarks the selection is purely arbitrary, and that is the case, up to a point. The "point" is that one particular distribution provides the most straightforward transformation to be used on any arbitrary normal distribution. That distribution is the *standardized* or *standard* normal distribution, a normal distribution with mean zero and variance 1.

Suppose we wish to find probabilities of a $N(\mu, \sigma^2)$ distribution where μ and σ^2 are arbitrary. If the random variable in question is x, then consider the following transformation of x:

$$z = \frac{x - \mu}{\sigma}. \tag{4A.1}$$

Taking $E(z)$ we find

$$\begin{aligned}
\mu_z &= E\left(\frac{x-\mu}{\sigma}\right) = \frac{1}{\sigma} E(x - \mu) = \frac{1}{\sigma} [E(x) - \mu] \\
&= \frac{1}{\sigma} (\mu - \mu) = 0, \\
\sigma_z^2 &= E[z - E(z)]^2 \\
&= E(z^2) - (0)^2 \\
&= E\left(\frac{x-\mu}{\sigma}\right)^2 \\
&= \frac{1}{\sigma^2} E(x - \mu)^2 \\
&= \frac{\sigma^2}{\sigma^2} = 1.
\end{aligned} \tag{4A.2}$$

Thus the transformation produces a new variable, z, with zero mean and unit variance.

To see that x is, in fact, also a normally distributed variable, we may substitute for x in terms of z in the density function $f(x)$ given by

$$f(x) = \frac{1}{\sqrt{2\pi}\,\sigma} e^{-(1/2)(x-\mu)^2/\sigma^2}, \tag{4A.3}$$

for the calculation of an arbitrary probability, say $P(a \leq x \leq b)$. The transformation is completed by transforming the limits (a, b) by equation (4A.1), substituting $x = z\sigma + \mu$ for x in (4A.3), and finally noting that dx in the original form is equivalent to $\sigma\, dz$ for the transformed variable. Thus equation (4A.3) becomes

$$P(a \leq x \leq b) = \int_{[(a-\mu)/\sigma]}^{[(b-\mu)/\sigma)]} \frac{1}{\sqrt{2\pi}} e^{-(1/2)z^2}\, dz, \tag{4A.4}$$

which shows the density function for z to be

$$f(z) = \frac{1}{\sqrt{2\pi}} e^{-(1/2)z^2}, \tag{4A.5}$$

corresponding to a normal distribution with mean zero and variance 1.

As an illustration of use of the $N(0, 1)$ tables, suppose we require $P(5 \leq x \leq 8)$ for $x \sim N(6, 25)$, as in Figure 4A-2. Transforming the boundary points $x = 5$ and

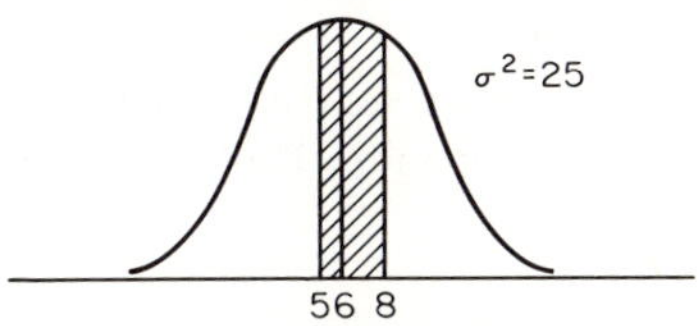

Figure 4A-2.

$x = 8$ to standard form gives $z_1 = (5 - 6)/5 = -\frac{1}{5}$ as the lower value, and $z_2 = (8 - 6)/5 = \frac{2}{5}$ as the upper value, as in Figure 4A-3.

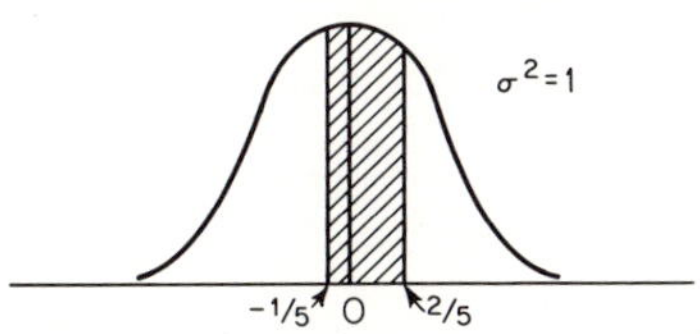

Using Appendix Table 2 we obtain $P(z \leq .40)$ as .6554. $P(z \leq -.20)$ is not available directly, since only cumulants for positive values of z are tabled. But since the normal is a symmetric distribution, the following identity holds: $P(z \leq -.20) = P(z \geq .20)$. This latter probability is obtained as the complement of $P(z \leq .20)$; i.e., $P(z \geq .20) = 1 - P(z \leq .20) = 1 - .5793 = .4207$ from the tables. The required probability is the difference between these two probabilities; i.e.,

$$P[5 \leq x \leq 8 \mid x \sim N(6, 25)] = P[-.20 \leq z \leq .40 \mid z \sim N(0, 1)]$$

$$= P[z \leq .40 \mid z \sim N(0, 1)] - P[z \leq -.20 \mid z \sim N(0, 1)] = .6554 - .4207 = .2347.$$

To approximate a binomial distribution, the procedure is to calculate the mean and variance for the particular binomial distribution in question and next to transform the limits in the required probability statement to standard form. The actual limits are first modified to account for the differences in form for the two distributions, i.e., discrete vs. continuous. The probability required is then obtained from the tables by a procedure such as the one outlined above.

Suppose the binomial distribution in question is $y \sim B(20, .5)$, so that $\mu_y = np = 10$ and $\sigma_y^2 = npq = 5$. Then applying the transformation with the appropriate "continuity corrections" to find $P(6 \leq y \leq 12)$, we get

$$z_1 = \frac{6 - \frac{1}{2} - 10}{2.236} = -1.56 \qquad z_2 = \frac{12 + \frac{1}{2} - 10}{2.236} = .67 \tag{4A.6}$$

The "sense" of the corrections applied obtains from splitting the area under the normal over any unit interval so that the approximation to a *given* binomial probability is

$$P[(x - \tfrac{1}{2}) \leq x \leq (x + \tfrac{1}{2}) \mid N(np, npq)] \simeq P[x \mid B(n, p)].$$

Using the standard normal tables we may calculate the required approximate probability as .6892. Referring to Appendix Table 1, the true binomial probability is .6906.

5

Sampling and estimation

With the previous three " background " chapters completed, we are in position to move into the main subject matter of this book, statistical inference. As mentioned previously, this subject may be discussed in two ways, via sample estimates of parameters and via the analysis of decision rule formation.

The topic treated in this chapter is the former, statistical inference about parameters from sample information, in the form of estimates of these parameters. The methodology is available for making *inferences* defined in terms of probability statements about the unknown parameters from their sample estimates. Thus we require knowledge of the frequency distribution of an estimator to make inferential statements about the parameter in question.

As an opening illustration of the fundamental ideas underlying statistical inference, let us take under consideration the following situation. A probability sample is taken of a well-defined group of doctors to discover the smoking habits of the whole group. Doctors are selected into the sample of n by a procedure which assigns an equal probability to the selection of any given doctor into the sample at any stage of the sampling procedure. Let us assume that a doctor is selected, answers yes or no to a question designed to register him either as a "smoker" or a "nonsmoker," and then is returned to the group. The same process is repeated until the entire sample has been gathered.

In the main group of doctors, the *population* for sampling purposes, there must be an actual proportion of "smokers" by the adopted definition, whatever it is. Call this proportion $\tilde{P}$. Now suppose we use the sample proportion, p, as an estimate of this unknown, $\tilde{P}$. That is, if of 20 doctors sampled we find 5 smokers, we estimate $\tilde{P}$ as $5/20 = .25$. Of course, $\tilde{P}$ either is or is not .25, in fact. But we, viewing the comparison ex ante, can never *know* the true relation between $\tilde{P}$ and any p. We can only discuss this relation in probability terms, expressing, for example, our "confidence" in p as an estimate of $\tilde{P}$.

When will we be more confident of p and when will we not? One way of using the term confidence is to relate it to the precision of p as compared to $\tilde{P}$. For example, since any *given* sample is a result of a random experiment and a *given* sample yields a particular value of p, we may conceptually generate a frequency distribution for p merely by considering all possible samples of a given size, n, as a sample space. The selection procedure assigns probabilities to their occurrence, and by the definition of the random variable p over this sample space we can construct $f(p)$. Presumably $\tilde{P}$ will be contained in the range of p; thus we may speak of the probabilities that a sample p will fall in various intervals from information about $f(p)$; for example, $P(p \geq \tilde{P})$, or $P(p \leq \tilde{P})$, etc. Of course, as a practical matter, to make any such probability statements we must know what parameters are involved in $f(p)$.

Precision may also be defined as we did in Chapter 1, simply as how "close" p is to $\tilde{P}$. Since p is a random variable we cannot, however, predict how "close" any particular sample proportion will be to $\tilde{P}$. This impasse leads to alternative ways of describing precision, the most conventional being through the second moment of p around $\tilde{P}$, called the mean-square-error of p: $\text{MSE}_p = E(p - \tilde{P})^2$. If it so happens that p has as its expected value $\tilde{P}$, then the above formula is just the variance of p. Again, we cannot know this variance since it involves $\tilde{P}$, the parameter we are attempting to estimate. Apparently we must also estimate σ_p^2 before we can say much about p as an estimate of $\tilde{P}$.

So, for example, we might devise a way to obtain an estimate of σ_p^2 and be able to quote as follows: that our estimate p of $\tilde{P}$ is .25 with a (estimated) standard deviation from $\tilde{P}$ of .05. This use of σ_p as the indicator of precision is, to reiterate, based on $E(p) = \tilde{P}$ holding. We may be more "confident" of this result than, for example, $p = .23$ with a standard deviation of .10.

Quite clearly we require the development of concepts of "confidence," "precision," etc., into a formal framework of analysis. We begin by treating the topic of random sampling as it pertains to the generation of frequency distributions for sample estimators of population characteristics.

1. Simple random sampling—finite population mean and variance

We may broadly dichotomize sampling procedures into probability methods and nonprobability or judgment methods. A *probability sample* guarantees that each member of a population of interest has a nonzero, known probability of selection into the sample, whereas a judgment method does not. Statistical inference of the sort we are herein concerned with requires probability sampling, and so we shall limit our discussion only to those methods. In particular we consider simple random sampling.

There exist many probability methods for sample selection, each of which may be "best" for some application. Of these the most basic is *simple random sampling*, where the probability that any sample of size n is selected is equal over all such samples. For example, for a population of size N and a sample of size n, the total number of different samples is just $\binom{N}{n}$ if each time an element of the population is selected it is not returned to the population prior to the next such selection. Then SRS is defined to be a technique of selection of the individual elements which gives $1\Big/\binom{N}{n}$ as the probability that a particular sample of n is drawn.

If individual elements are *replaced* into the population (i.e., there exists some chance that the *same* element will be drawn n times), then there are N^n possible size n samples, each of which must be guaranteed a probability $1/N^n$ under SRS.

The selection method which ensures these conditions are met in the two cases outlined is called an *EPSEM*, Equal Probability Selection Method. In the first case, where sampling is done *without* replacement, at *each stage* in the selection procedure the *remaining* population elements are assigned† equal chances of being selected on the next draw. Without elaborating, it is easy to argue that each possible sample of size n must be equally likely to obtain. For sampling with replacement any particular element has an equal chance of being selected at any stage of the drawing process, and, indeed,

† Through a random-numbers table or other suitable device.

this probability remains constant throughout the drawings. Again, it is quite clear that each possible n-sample will be equally likely, with probability $1/N^n$.

What is the probability that a given n-sample will contain a particular population element? Letting $A_1, \ldots, A_N$ be the elements, under sampling with replacement the probability in question may be restated as: "What is the probability that an n-sample contains A_i at least once?" As with the birthday problem of Chapter 2, it is easier to find the complementary probability that A_i will not appear. This is just $[(N-1)/N]^n = [1-(1/N)]^n \simeq 1-(n/N)$.† Thus the probability that A_i is contained in a given n-sample under sampling with replacement is approximately n/N. For sampling without replacement, the probability that A_i is *not* contained in our n-sample is just the number of ways we can select an n-sample from the other $(N-1)$ elements divided by the total number of n-samples from N elements; i.e.,

$$P(A_i \textit{ not } \text{contained}) = \frac{\binom{N-1}{n}}{\binom{N}{n}} = 1 - \frac{n}{N}, \tag{1.1}$$

so that the result of interest is the same: The probability that the element A_i is in a given n-sample is n/N.

These, then, are the attributes of samples selected by EPSEM procedures and the SRS method. Other probability-oriented techniques are available and in many instances are more efficient than SRS in collecting sample information, where efficiency is defined as yielding information of a given precision at lowest cost. But to examine principles of inference an in-depth excursion into sampling theory is unnecessary, and so we shall confine ourselves to the two SRS techniques only.

The objective of sampling is to gain information about certain characteristics of the population $\{A_i\}$. What characteristics? For the population of doctors in a previous example, the "characteristic" was a population proportion. We might as well be concerned with other quantitative insights into the population, such as its mean value for some measurement taken on the $\{A_i\}$, a variance of this measured attribute, etc.

The simplest way to deal with these attributes of a population is to view the population elements as elementary events in a sample space $\{A_i\}$. Suppose we assign equal probability weights to these A_i's and define the measured (or qualitative) attribute of interest as a random variable, say $\overline{\underline{X}}$. By this contrived set of conditions we must be able to produce $f(\overline{\underline{X}})$ a frequency function for the attribute, defined via the population. This $f(\overline{\underline{X}})$

† If N is large relative to n so that terms of the order of smallness $1/N^2$ and smaller may be neglected.

by definition merely collects A_i's with equal values of $\underset{-}{X}$ and adds their probabilities. So, for example, the population might contain five elements which have the value $\underset{-}{X} = k$. By the assignment of equal weights to each A_i, $f(\underset{-}{X} = k) = 5/N$.†

Now $f(\underset{-}{X})$ may be used to define the population mean μ simply as its expected value; i.e.,

$$\mu = E(\underset{-}{X}) = \sum_{\text{all } \underset{-}{X}} \underset{-}{X} f(\underset{-}{X}) = \frac{1}{N} \sum_{i=1}^{N} X_i, \tag{1.2}$$

where X_i is the measured attribute on A_i. Similarly, a population variance σ^2 may be obtained as

$$\begin{aligned} \sigma^2 &= E(\underset{-}{X} - \mu)^2 \\ &= \frac{1}{N} \sum_{i=1}^{N} (X_i - \mu)^2 \\ &= \frac{1}{N} \sum_{i=1}^{N} X_i^2 - \mu^2. \end{aligned} \tag{1.3}$$

If a qualitative, dichotomous attribute is of interest, then let $X_i = 1$ if A_i possesses the attribute of interest and $X_i = 0$ if not. Then (1.2) counts the number of A_i which possess the attribute, and μ defines the population proportion $\tilde{P}$.

For the purposes of this book we shall concern ourselves only with sample inferences about these population characteristics, its mean (or a proportion) and variance. Of course there are many other characteristics which might be of interest, but in principle the methodology we shall propose may be extended to cover any such problem.

2. Estimates and their properties

Suppose we have a simple random sample of size n. The "information" contained in the sample takes the form, obviously, of the measurements X_j on the sample subset of A_j. If we wished to say something—infer something—about μ, for instance, how should we utilize these sample observations? One very intuitive "estimator"‡ of μ is the sample arithmetic average, say $\bar{x}_n$, where

† Here we are dealing with finite populations, as we have implicitly from the beginning of this chapter; we shall extend our results to the infinite population case later.

‡ An *estimator* is the functional form in which sample observations are utilized; an *estimate* is the value the estimator takes on for a particular set of observations.

$$\bar{x}_n = \frac{1}{n} \sum_{j=1}^{n} X_j. \tag{2.1}$$

The problem, then, is to evaluate the relationship of $\bar{x}_n$ and μ in probability terms. For there is, from the definition of the SRS method, a definite probability attached to the obtention of this sample. Therefore, looking at all possible samples of this same size, n, $\bar{x}_n$ is a random variable with a frequency function $f(\bar{x}_n)$. A sample of size 20 from a population of 500 which yields $\bar{x}_{20} = 5u$ (units) *may* have been generated by $\mu = 5u$, $\mu = 16u$, $\mu = 103u$, etc. Can we provide an objective judgment as to which μ is the true one? Perhaps this goal may be accomplished via a statement of the likely interval in which μ is contained. In any case the objective judgment we refer to cannot be a judgment with certainty since samples are, in general, smaller than the entire population.†

One such evaluation which appears useful a priori is the position of μ in $f(\bar{x}_n)$. Specifically, is $E(\bar{x}_n) = \mu$? We already (by definition) know that μ is the mean of the population; is it also the mean of the *sampling distribution* of $\bar{x}_n$? Given that we can specify $f(\bar{x}_n)$ we may answer the question directly by merely finding $E(\bar{x}_n)$.

For the case at hand, $\bar{x}_n$ is a linear combination of the X_j. Ex ante we cannot predict which A_k will be selected to fill the next vacant slot in the sample, and hence we cannot predict the *value* X_k. In fact, the value which does (ex post) fill any given slot is the result of a random selection. Let us first consider the initial sample value, X_1. The frequency distribution which governs the actual value X_1 is $f(\underline{\overline{\mathrm{X}}})$. So, if we selected a sample of size $n = 1$ only, it is easy to see that $E(\bar{x}_1) = \mu$.

For notational purposes, let us define x_j as the random variable corresponding to the jth selection into the sample, with frequency distribution $f(x_j)$, and consider the extension of the analysis above to $n = 2$ under the assumption that sampling is done *with replacement*. Then x_1 and x_2 are independent random variables and also $f(x_1) = f(x_2) = f(\underline{\overline{\mathrm{X}}})$. Then by rule 1 for expected values,

$$\begin{aligned} E(\bar{x}_2) &= \tfrac{1}{2}E(x_1 + x_2) \\ &= \tfrac{1}{2}[E(x_1) + E(x_2)] \\ &= \tfrac{1}{2}(2\mu) \\ &= \mu. \end{aligned} \tag{2.2}$$

† It may not even be possible to identify parameters exactly, or, in different words, make certain judgments about them *given* the entire population. See Chapter 7 for some elaboration on this point.

Under these assumptions, the mean of the sample means for 2-samples is μ, the population mean. The extension of (2.2) to n-samples should be obvious. In that case also we find that $E(\bar{x}_n) = \mu$.

What if sampling is done without replacement? Clearly the $\{x_j\}$ are *not* independent.† Thus theoretically the $\{x_j\}$ do not possess identical frequency functions, as they did before. If N is quite large in relation to n, however, the fact of nonindependence will not bear materially on the previous results, as the $f(x_j)$ may be considered identical empirically.

In general terms the property that an estimator have the parameter of interest as its expected value in each sampling distribution (i.e., independent of the actual sample size) is called *unbiasedness*. If b_n is an estimator of the population value B for an n-sample, unbiasedness implies, for example, the graphs of Figure 5-1. The definition is: For b_n an estimator of B, if $E(b_n) = B$, then b_n is said to be an *unbiased* estimator of B.

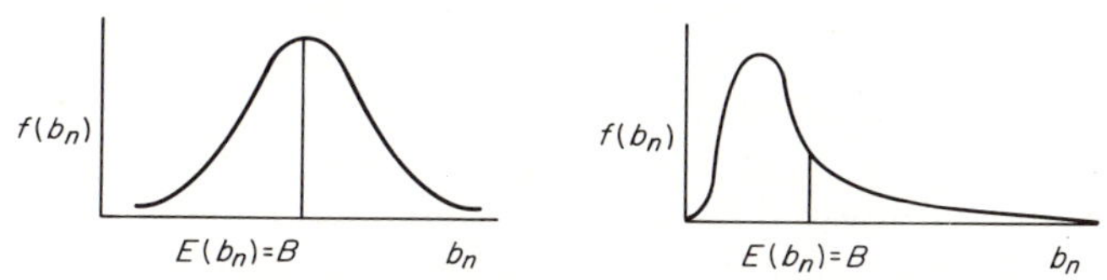

Figure 5-1. Examples of unbiased estimation.

If this is the definition of unbiased estimation, what is *biased* estimation? Very simply it is the opposite of unbiased estimation, i.e., when $E(b_n) \neq B$. Figure 5-2 depicts a situation of biased estimation. The quantity $E(b_n) - B$

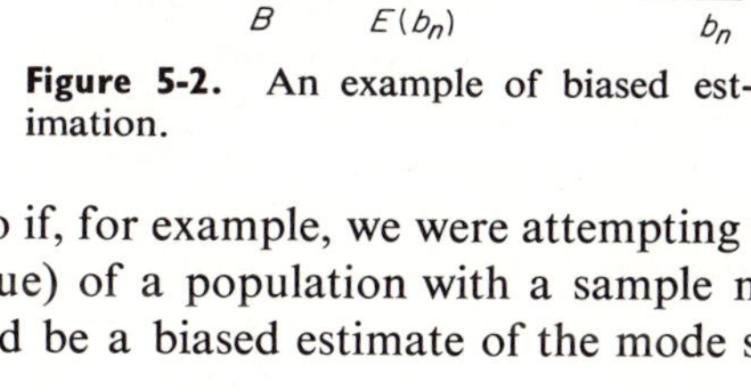

Figure 5-2. An example of biased estimation.

is called the *bias*. So if, for example, we were attempting to estimate the *mode* (most frequent value) of a population with a sample mean, $\bar{x}_n$, which has $E(\bar{x}_n) = \mu$, $\bar{x}_n$ would be a biased estimate of the mode so long as $f(\bar{x}_n)$ was asymmetric.‡

† Because their joint frequency function $f(x_1, \ldots, x_n)$ is not the product of their respective individual functions $f(x_1), \ldots, f(x_n)$.

‡ We are assuming here that $f(\bar{x}_n)$ is unimodal.

To illustrate these concepts, let us elaborate on the opening example of this chapter, concerning the estimation of the proportion of smoking doctors, $\tilde{P}$. Again, an intuitive estimator of $\tilde{P}$ is p_n, the sample proportion for n observations. If we put $X_k = 1$ when the observed element A_k is a smoker and $X_k = 0$ if not, then the sample proportion, p_n, takes the general form of an arithmetic mean; i.e.,

$$p_n = \frac{1}{n} \sum_{j=1}^{n} x_j. \tag{2.3}$$

From our discussion of Section 1, if sampling is done with replacement so that $f(x_1) = f(x_2) = \cdots = f(x_n)$, then the expected value of p_n is

$$\begin{aligned} E(p_n) &= \frac{1}{n} \sum_{j=1}^{n} E(x_j) \\ &= \tilde{P}, \end{aligned} \tag{2.4}$$

since each x_j has the same expectation, $\tilde{P}$.

Note that $np_n = \sum_{j=1}^{n} x_j$ has the form of a binomial random variable. Indeed, we may deduce that sampling for dichotomous attributes with replacement must theoretically produce a binomial sampling distribution for np_n, since each slot to be filled in the sample is the analogue for a Bernoulli trial, where the probability of success is just the probability that the chosen element possesses the attribute of interest. This probability is $\tilde{P}$. Thus the expected value of np_n, a binomial random variable, is $n\tilde{P}$ as before; the expected value of p_n is obviously $\tilde{P}$, and p_n is *also* governed by a binomial distribution since when $np_n = k$, $p_n = k/n$, etc. The only difference, then, between the distribution of np_n and p_n is that the abscissa values of the distribution of np_n are divided by n.

If sampling is accomplished without replacement, the probability of "success" is *not* constant over the n drawings (trials); the trials are *not* independent. The probability that the first element selected (at random) possesses the attribute is $\tilde{P} = Y/N$, where Y is the number of elements possessing the attribute in the population. But the probability of success on the second draw clearly depends on what happens on the first. If an element with the attribute is selected first, then $(Y - 1)/(N - 1)$ is the probability that a similar occurrence applies to the second draw. But if no attribute was observed on draw 1, then this probability is $Y/(N - 1)$, etc. If we generalize this process, we find† the distribution of np_n to be

$$P(np_n = k \mid N, Y, n) = \frac{\binom{Y}{k}\binom{N-Y}{n-k}}{\binom{N}{n}}, \tag{2.5}$$

† See the Appendix to this chapter.

which is the general term of the *hypergeometric* distribution and bears a close resemblance to our calculation techniques for obtaining probabilities that card hands will contain certain configurations of cards.† Indeed, when we wish to obtain probabilities in problems where the population (52 cards, perhaps) is dichotomized, the hypergeometric distribution is the governing distribution when sampling is accomplished without replacement.

What can be said about p_n as an unbiased estimate of $\tilde{P}$ when sampling is done without replacement—when the hypergeometric distribution is the sampling distribution? Theoretically, for any $n < N$, we cannot say much unless we assume the size of N, Y, and n are such that the differences between, e.g., $Y/(N-1)$, $(Y-1)/(N-1)$, etc., are very small and may, therefore, be ignored empirically. Such an assumption is tantamount to assuming that the probability of success on each trial is a constant, $\tilde{P}$. Thus the previous binomial case holds for which unbiasedness has already been shown. It is *not necessary* to make such an assumption for purposes of inference about Y, the number of population elements that possess the attribute or $\tilde{P}$. We only use it here to give some intuitive feeling for the relationship between the hypergeometric and binomial distributions and to provide a simple (however, restrictive) empirical indication of the unbiasedness of a sample proportion in the case of sampling without replacement.‡

Obviously there are many ways one can combine sample observations for the purpose of estimating μ, only one of which is as an arithmetic mean, $\bar{x}_n$. $\bar{x}_n$ does possess the desirable property of unbiasedness as an estimator of μ, but that may not be a sufficient criterion to use in deciding which, among many, estimators to use as an estimator of μ or, in general, of any given population characteristic.

Consider, for example, $y_n = \bar{x}_n + (1/n)$ as an estimator of μ. Upon taking its expectation we see that $E[\bar{x}_n + (1/n)] = \mu + (1/n)$, so that y_n is a biased estimator of μ, the bias being $1/n$. Now if n is "large," it is apparent that $y_n \simeq \bar{x}_n$, and under sampling with replacement $y_n \to \bar{x}_n$ as $n \to \infty$, which by our previous work implies $E(y_n) \to \mu$ as $n \to \infty$. This *asymptotic unbiasedness* is called the property of *consistency* if in conjunction with $E(y_n) \to \mu$ there is an accompanying clustering of values y_n around μ. This provides a "loose" definition of consistency. If this latter condition did not hold, then the bias would not *progressively* move toward zero; i.e., there would not necessarily be any benefit toward better inference from taking larger samples, a relationship we perhaps take for granted a priori.

Under what circumstances might a biased estimator be preferred to an unbiased one? Also, how may we reasonably choose between two unbiased estimators? The latter question will be dealt with first, and finds its answer in

† Cf. the Appendix to Chapter 2.

‡ As a practical matter the binomial provides a useful approximation to the hypergeometric, and we shall assume throughout the book that this approximation is a "satisfactory" one. (See Chapter 7 for a definition of the term "satisfactory" as we use it here.)

a comparison of the precision of alternative estimators as measured by their respective second moments around the population constant. Clearly an estimator which has its possible values (possible over attainable samples of a given size) clustered around the population parameter is more desirable than one whose values (for the same sample size) are more widely dispersed, simply because the probability that any given n-sample will produce an estimate in some arbitrary range is larger in the former case. This probability is some measure of our "confidence" in the estimate we actually obtain, but is obtained by ex ante considerations.

Consider the comparison indicated in Figure 5-3, where the distributions

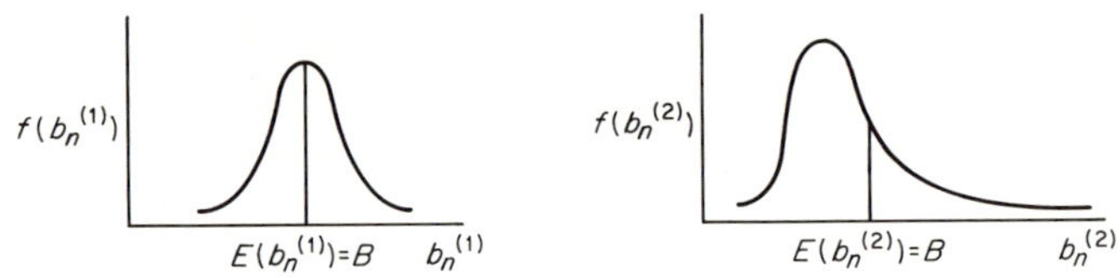

Figure 5-3. Comparison of precision for two unbiased estimators.

of two estimators $b_n^{(1)}$ and $b_n^{(2)}$ of B are shown. Although both estimators have B as their expected value, $b_n^{(1)}$ appears to be superior, since the probability that any actual $b_n^{(1)}$ falls within some small interval to either side of B is larger than if $b_n^{(2)}$ were used. To summarize this criterion we may use the *mean-square-error* concept. We define the MSE for any estimator as follows:

$$\text{MSE}_{b_n} = E(b_n - B)^2, \tag{2.6}$$

an indicator of how the actual values of an estimator are distributed around the constant B.

Some manipulation of (2.6) gives MSE in terms of the variance of b_n, $\sigma_{b_n}^2$, and the bias $[E(b_n) - B]$: Adding and subtracting $E(b_n)$ in (2.6),

$$\begin{aligned} \text{MSE}_{b_n} &= E[b_n - E(b_n) + E(b_n) - B]^2 \\ &= \sigma_{b_n}^2 + [E(b_n) - B]^2 + 2E[b_n - E(b_n)][E(b_n) - B]. \end{aligned} \tag{2.7}$$

Now the last term may be rewritten $2[E(b_n) - B]E[b_n - E(b_n)]$, since $E(b_n) - B =$ bias is a constant. But $E[b_n - E(b_n)] = 0$, so

$$\text{MSE}_{b_n} = \sigma_{b_n}^2 + [E(b_n) - B]^2. \tag{2.8}$$

For a statement of the MSE criterion we shall say that an estimator $b_n^{(1)}$ of B is more *efficient* than $b_n^{(2)}$ if $\text{MSE}_{b_n}(1) < \text{MSE}_{b_n}(2)$. Note that if b_n is an unbiased estimator, (2.8) becomes

$$\text{MSE}_{b_n} = \sigma_{b_n}^2 \qquad \text{if } E(b_n) = B. \tag{2.9}$$

Then for purposes of selecting a "best" estimator from among several *unbiased* alternatives, the MSE criterion (in the case, a variance criterion) provides an unambiguous rule. Likewise, if the alternatives under consideration are all biased estimators with *equal* biases for a given sample size, the MSE criterion will select a "best" estimator. If, however, the choice is to be made from among a set of estimators with varying attributes the chosen estimator will balance large variance with small bias or vice versa.

Returning to the comparison of $\bar{x}_n$ and $\bar{x}_n + (1/n)$ as estimators of μ, we know $E(\bar{x}_n) = \mu$ but $E[\bar{x}_n + (1/n)] = \mu + (1/n)$. Applying the MSE criterion directs us to compare $\text{MSE}_{\bar{x}_n} = \sigma^2_{\bar{x}_n}$ with $\text{MSE}_{\bar{x}_n+(1/n)} = \sigma^2_{\bar{x}_n} + 1/n^2$. In this case the unbiased estimate also possesses smallest MSE and so is judged superior.

As a further hypothetical illustration of these ideas, suppose another competitor to $\bar{x}_n$ as an estimator of μ is given by

$$\tilde{x}_n = \sum_{j=1}^{n} w_j x_j \qquad \sum_{j=1}^{n} w_j = 1, \tag{2.10}$$

where the x_j's are random variables as defined previously. It is easy to see that

$$E(\tilde{x}_n) = \sum_{j=1}^{n} w_j E(x_j) = \mu \sum_{j=1}^{n} w_j = \mu. \tag{2.11}$$

Thus, if we specify any set of w_j so that $\sum w_j = 1$ and combine the sample observations by (2.10), we shall produce an estimate with an expected value of μ. Suppose for $n = 3$ we specify $w_1 = \frac{1}{2}$, $w_2 = \frac{1}{4}$, $w_3 = \frac{1}{4}$, and observe $x_1 = 5u$ (units), $x_2 = 3u$, and $x_3 = 10u$. Then $\bar{x}_3 = 18u/3 = 6u$, and $\tilde{x}_3 = 5\frac{3}{4}\,u$. Both estimators are unbiased: If we selected all possible samples of size 3 and calculated $\bar{x}_3$ and $\tilde{x}_3$, then averaged these values over all samples, we would find that *both* averages equaled μ. Upon calculating the variances for each (since they are unbiased), however, we would find that $\bar{x}_3$ is the more efficient estimator.

In general, if the sample-draw random variables are independent, and have the same $f(x_j)$, with $E(x_j) = \mu$ and $\sigma^2_{x_j} = \sigma^2$ (the population variance), then

$$\sigma^2_{\bar{x}_n} = \frac{1}{n^2} \sum_{j=1}^{n} \sigma^2_{x_j} = \frac{n}{n^2} \sigma^2 = \frac{\sigma^2}{n}, \tag{2.12}$$

using rules 1 and 2 for variances. This formula applies under sampling with replacement. Using the same assumptions for $\tilde{x}_n$ we find

$$\sigma^2_{\tilde{x}_n} = \sum_{j=1}^{n} w_j^2 \sigma^2_{x_j} = \sigma^2 \sum_{j=1}^{n} w_j^2. \tag{2.13}$$

Whether $\bar{x}_n$ is a more efficient estimator depends on $1/n < \sum_{j=1}^{n} w_j^2$, which holds except when all $w_j = 1/n$, i.e., when $\tilde{x}_n = \bar{x}_n$.† Thus the sample mean and, therefore, the sample proportion are "best" estimators of μ and $\tilde{P}$ among all linear estimators, those estimators which appear as linear functions of the sample observations.

As pointed out previously, in actuality choice among alternative estimators in some cases may involve unbiased versus biased estimators when the biased estimator possesses a smaller MSE. Thus, although for a given sample size the unbiased estimator has the population parameter as its average value, it may be relatively imprecise compared to the biased estimator. Which estimator to use in this specific case depends on the user's judgment, but a conventional rule is to choose the estimator with smallest MSE (i.e., the most efficient), which implies an equal distaste for bias (squared) and dispersion (as measured by the variance).

As a final note in this discussion of properties of estimators in general, one need not rely on intuition or experience to produce alternative estimators prior to an evaluation via these criteria. There are techniques available for generating estimators which already possess certain properties, notably the methods of least squares and maximum likelihood. LS, to be applicable, depends on certain assumptions being met but produces estimates of minimum MSE. The maximum likelihood method requires and utilizes the joint frequency distribution of sample observations and produces consistent estimators. The techniques rest on different assumptions but in some cases may both be applicable, in which instance LS is the superior method, usually due to the efficiency of the estimator produced and the lack of required knowledge of the distribution of the estimator (in the form of the distribution of sample observations).

For while it may be easy to justify the binomial or some other theoretical distribution of certain situations in sampling, it is not possible in most others to argue that some distribution a priori generated the sample. Here we are speaking mainly of the lack of knowledge about the *form* of the sampling distribution rather than parameters given the form. The central limit theorem, which was touched upon before, may be invoked to argue that the normal distribution is a reasonable approximation to the true

† This inequality is easy to demonstrate with the aid of calculus. We wish to find the set of w_j which minimize $\sum_{j=1}^{n} w_j^2$ subject to the restriction that $\sum_{j=1}^{n} w_j = 1$. The solutions are available using the Lagrangian multiplier technique from the equations

$$\begin{array}{ll} j = 1, \ldots, n & 2w_j + \lambda = 0 \\ n + 1 & \sum w_j - 1 = 0 \end{array}$$

Every w_j must equal $-\lambda/2$ and $-n\lambda/2 = 1$, so $\lambda = -2/n$ and $w_j = 1/n$ ($j = 1, \ldots, n$). This minimization is an example of the method of least-squares estimation, where we attempt to deduce the estimator (a function of sample observations) which provides minimum variance (MSE) over the n-sample.

distribution of some estimators.† However, it is difficult to assess what "reasonableness" is, and then to know when sample size is large enough for the approximation to be satisfactory.

Reflecting on our introductory remarks on inference concerning the estimate-parameter relationship, this lack of knowledge concerning the form of an estimator's sampling distribution appears as a serious obstacle. To provide probability statements about this relationship we must be able to calculate the probabilities. Fortunately, however, we may avert some of these problems through the use of a famous theorem of statistics, *Tchebycheff's inequality*.

3. Tchebycheff's inequality—confidence intervals

Tchebycheff's inequality allows us to make probability statements about an estimator and how it is distributed around a population constant *without* knowledge of the form of this distribution. Thus it provides a theoretical basis for such statements when any particular distribution is assumed or justified and is of great generality. The theorem may be easily stated and proved:

Let x be a random variable and k a constant. Then for *any* $e > 0$,

$$\text{(a)} \qquad P(|x-k| > e) < \frac{\sigma_x^2 + [E(x)-k]^2}{e^2}, \qquad (3.1)$$

and since $P(|x-k| > e) + P(|x-k| \le e) = 1$,

$$\text{(b)} \qquad P(|x-k| \le e) \ge 1 - \frac{\sigma_x^2 + [E(x)-k]^2}{e^2}.$$

The numerator of the fraction is simply MSE_x, and e is any positive number representing a preset bound on the absolute difference between x and k.

If x is an unbiased estimator of k, then $E(x) = k$, and a special case arises:

$$\text{(a)} \qquad P(|x-k| > e) < \frac{\sigma_x^2}{e^2}, \qquad (3.2)$$

$$\text{(b)} \qquad P(|x-k| \le e) \ge 1 - \frac{\sigma_x^2}{e^2}.$$

We shall proceed to prove (a), which implies (b). The proof is in terms of a discrete random variable for convenience only. The result applies equally well to continuous random variables.

† Section 6, Chapter 3. The estimator in question must possess a finite variance and must, in general, exhibit consistency for the theorem to be applicable.

Let x have possible values $x_1, \ldots, x_n$ with probabilities $p_1, \ldots, p_n$. Suppose that of the n differences $(x_i - k)$, exactly m of them are such that $|x_i - k| > e$, and, therefore, $n - m$ of them have $|x_i - k| \leq e$. We may easily speak of the probability $P(|x - k| > e)$ in terms of the m possible values of x which have $|x_i - k| > e$. That is,

$$P(|x - k| > e) = \sum_{i=1}^{m} p_i.$$

Consider the expected value of $(x - k)^2$, given by

$$E(x - k)^2 = \sum_{i=1}^{n} (x_i - k)^2 p_i.$$

It follows directly that

$$E(x - k)^2 \geq \sum_{i=1}^{m} (x_i - k)^2 p_i.$$

Now if $|x_i - k| > e$, then $(x_i - k)^2 > e^2$, so that

$$E(x - k)^2 > \sum_{i=1}^{m} e^2 p_i = e^2 \sum_{i=1}^{m} p_i,$$

$$\frac{E(x - k)^2}{e^2} > \sum_{i=1}^{m} p_i.$$

But from our initial specification, $\sum_{i=1}^{m} p_i = P(|x - k| > e)$, which leads to the final result,

$$P(|x - k| > e) < \frac{E(x - k)^2}{e^2}.$$

The above probability statement may, of course, be rewritten without absolute-value notation. Form (b) is more commonly used, however, and when the absolute-value signs are removed appears as

$$P(x - e \leq k \leq x + e) \geq 1 - \frac{E(x - k)^2}{e^2}. \tag{3.3}$$

Equation (3.3) is a classical statement of the *confidence interval*: If x is an estimator of k, then the "probability that the interval $(x - e, x + e)$ contains k is ...."

To relate the confidence interval to our previous notions of the precision of an estimate, note that the value of e may be arbitrarily set, say as so many root-mean-square error units, etc.† Suppose, for example, x is an unbaised estimator of k. Then the special case of the theorem applies, and $E(x - k)^2 = \sigma_x^2$. Also, let $e = 3\sigma_x$. That is, we are requesting the probability that $x \pm 3\sigma_x$ will contain the parameter k. The probability is

$$P(x - 3\sigma_x \leq k \leq x + 3\sigma_x) \geq 1 - \frac{\sigma_x^2}{9\sigma_x^2} = 1 - \frac{1}{9} \qquad (3.4)$$

$$\geq \tfrac{8}{9}.$$

Use of the theorem obviously requires *knowledge* of σ_x, which, in turn, depends on a complete knowledge of the sampling distribution of an estimator, information we cannot possess. (Otherwise, why would we sample?) And, to be sure, we only take *one* sample. How does this probability statement relate to the problem of interpretation for one sample?

An answer to the first question must wait briefly; the second may be treated as follows, from the definition of probability. Suppose 90 samples were taken at random, each of a given size, n. For each sample the estimate x (an actual, calculated value) is combined with the arbitrary constant $e = 3\sigma_x$ to form 90 intervals. The interpretation is that were these 90 intervals to be plotted as in Figure 5-4, they would enclose k with a relative frequency at

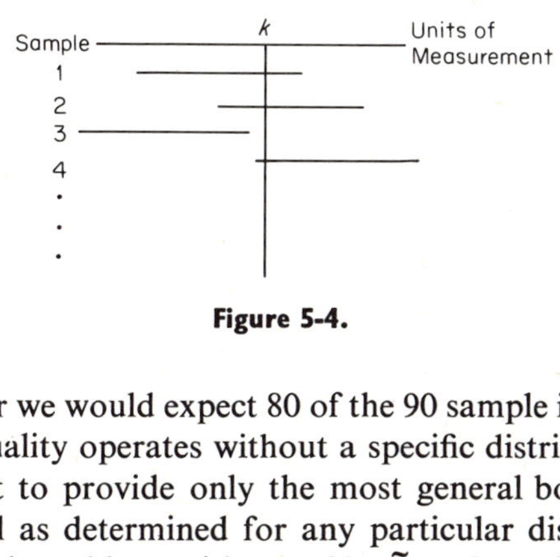

Figure 5-4.

least equal to 8/9, or we would expect 80 of the 90 sample intervals to cover k.

Since the inequality operates without a specific distribution assumption, we should expect it to provide only the most general boundaries on actual probabilities, actual as determined for any particular distribution.

For a binomial problem with $n = 30$, $\tilde{P} = .5$, and using the sample proportion, $p_n = y/n$ as an estimator of $\tilde{P}$ (unbiased), rule 2 for variances yields $\sigma_p^2 = (1/n^2)\sigma_y^2 = (n\tilde{P}\tilde{Q})/n^2 = (\tilde{P}\tilde{Q})/n = .25/30 = .008$. $\sigma_p = \sqrt{.008} \simeq .09$

† Although the bounds produced for, e.g., $e \leq 1\sqrt{\text{MSE}_x}$ are not very informative.

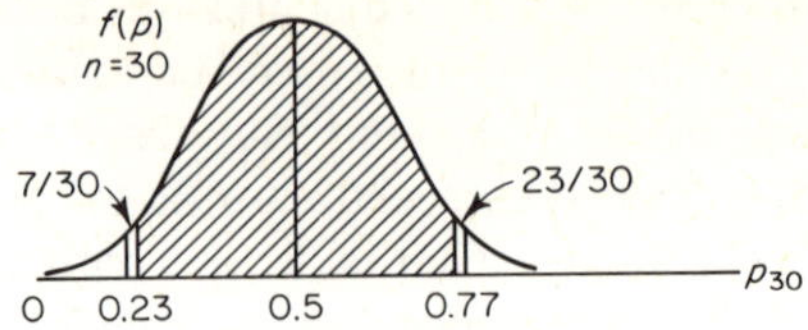

Figure 5-5.

and $3\sigma_p = .27$, corresponding to Figure 5-5, where a continuous function is drawn for visual ease only. The calculated 3σ limits are .23 and .77 with 7/30 and 23/30 the corresponding exact binomial limits.† The binomial probability is .9986, whereas Tchebycheff's inequality says the interval holds with probability *at least* .89 (8/9).

Suppose we were estimating μ with the mean of a sample of size 25 from a normally distributed population with $\sigma^2 = 100$. If the sample random variables are x_j, $j = 1, \ldots, n$, with $x_j \sim N(\mu, 100)$ for all j, then for the sample mean $\bar{x}_n = (1/n)\sum_{j=1}^{n} x_j$, $E(\bar{x}_n) = \mu$ [cf. equation (2.11)] and $\sigma^2_{\bar{x}_n} = \sigma^2/n$ [cf. equation (2.12)], so that $\bar{x}_n \sim N(\mu, \sigma^2/n)$ if sampling is accomplished by an SRS method.‡

For a confidence interval with $e = 2\sigma_{\bar{x}}$, Tchebycheff's inequality states the probability that μ is contained within the limits $\bar{x}_n \pm 2\sigma_{\bar{x}_n}$ is at least .75. The actual probability may easily be obtained from the standard normal tables as corresponding to the area under the $N(0, 1)$ distribution in the interval $(-2, 2)$; this probability is .9544.

It is easy to see how a probability statement from a normal sampling distribution about the (e.g.) sample mean being contained within fixed limits may be turned around to yield a statement about the fixed parameter being contained within variable limits. For the example just treated, the probability that a normally distributed n-sample mean is contained within the limits $\mu \pm 2\sigma_{\bar{x}_n}$ is just

$$P(\mu - 2\sigma_{\bar{x}_n} \leq \bar{x}_n \leq \mu + 2\sigma_{\bar{x}_n}) = .9544 \tag{3.5}$$

and corresponds to the shaded area of Figure 5-6. In actual units of measurement $2\sigma_{\bar{x}_n} = 2\sqrt{100/25} = 20/5 = 4$, so that the above probability is equivalent to

$$P(\mu - 4 \leq \bar{x}_n \leq \mu + 4) = .9544. \tag{3.6}$$

† The values .23 and .77 are *not* possible values, so that $P(.23 \leq p_{30} \leq .77) \equiv P(7/30 \leq p_{30} \leq 23/30)$.

‡ Cf. Appendix 2 to Chapter 4, and note that the (continuous) normal distribution can only be defined on a population of infinite size.

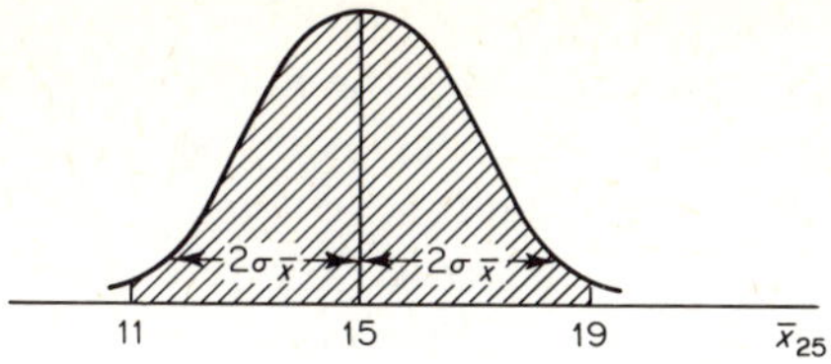

Figure 5-6.

In (3.5), if we operate on the inequalities, we can put $\bar{x}_n$ and μ in juxtaposition, producing the confidence interval as required. Merely add $-\bar{x}_n$ and $-\mu$ to all three sides of the inequality, to obtain

$$P(-\bar{x}_n - 2\sigma_{\bar{x}_n} \leq -\mu \leq -\bar{x}_n + 2\sigma_{\bar{x}_n}) = .9544. \tag{3.7}$$

Multiplying inside by (-1) and consequently reversing the inequalities,

$$P(\bar{x}_n + 2\sigma_{\bar{x}_n} \geq \mu \geq \bar{x}_n - 2\sigma_{\bar{x}_n}) = .9544, \tag{3.8}$$

which is identical to

$$P(\bar{x}_n - 2\sigma_{\bar{x}_n} \leq \mu \leq \bar{x}_n + 2\sigma_{\bar{x}_n}) = .9544. \tag{3.9}$$

For other distributions, such as the binomial and Poisson, the confidence interval or *interval estimate* of a parameter is not so easily described, and must wait for presentation later (Chapter 6). The reader should also note that the examples used have involved unbiased estimates. Tchebycheff's inequality may be utilized with e being quoted in terms of $\sqrt{\text{MSE}}$ units, but for the normal distribution the basic unit of measure is σ, the standard deviation of the distribution. Thus if the problem situation concerns a normally distributed, *biased* estimator, it is not true that the confidence interval may be quoted in $\sqrt{\text{MSE}}$ units of deviation to accomplish the required adjustment.† Our work will not deal with this problem, and we shall use unbiased estimators throughout (of population means), so that standard deviation units of measurement are appropriate in the normal case.

The relation between sample size and precision as expressed in the normal confidence interval is clear from (3.9). If we put $\sigma_{\bar{x}_n} = \sqrt{\sigma^2/n} = \sigma/\sqrt{n}$, then $n \to \infty \Rightarrow \sigma_{\bar{x}_n} \to 0$, which must hold since we are dealing with an unbiased (and consistent) estimator of μ. Thus the limits around μ may be tightened (more precise estimation) with increases in n at a given *confidence level.* Or, with given n the limits may be tightened by *lowering* the confidence level.

† See, for example, W. G. Cochran, *Sampling Techniques*, 2nd ed. (Wiley, 1963), pp. 12–16 for a clear demonstration of the problem.

In either case *it is assumed that* $\sigma^2_{\bar{x}_n}$ *(hence* σ^2*) is known*. Such knowledge is not available in most instances, but the normal interval depends upon it. Our methodology must, therefore, necessarily be modified to include such a situation, if possible. It would seem that to *know* σ^2 for a population implies knowledge of μ, which is indeed true. We return, therefore, to the now familiar circumstance of seemingly having to know what we are trying to estimate before we estimate it. The key to this obstacle is to note that σ^2 for a population may be estimated from a sample also. We turn then to an investigation of the effects of estimating σ^2 on the normal confidence interval.

4. Estimation of variance

The classical confidence interval for a normally distributed, unbiased estimator is given by

$$P(b_n - Z_{\alpha/2}\sigma_{b_n} \leq B \leq b_n + Z_{\alpha/2}\sigma_{b_n}) = 1 - \alpha, \tag{4.1}$$

where b_n is the n-sample estimator, B the population parameter of interest, σ_{b_n} the standard deviation of b_n around B, $1 - \alpha$ the confidence (probability) level, and $Z_{\alpha/2}$ the standard normal deviate which cuts off exactly $\alpha/2$ (proportion) of the area measured from the standard normal mean, zero, in each tail of the distribution. Our equation (3.5), for instance, is merely an application of this more general form, with $b_n = \bar{x}_n$, $B = \mu$, $1 - \alpha = .9544$, and $Z_{\alpha/2} = 2$.

What sort of confidence statement is available when $\sigma_{\bar{x}_n}$ is unknown? The answer to this question may be given only after we treat the more general problem of how to estimate $\sigma_{\bar{x}_n}$. Once this is accomplished we return to the confidence statement and attempt to analyze the effects of using an estimate (random variable) of $\sigma_{\bar{x}_n}$ rather than the parameter itself.

Equation (2.12) shows the key relationship between the variance of the sampling distribution of sampling means and the population variance,

$$\sigma^2_{\bar{x}_n} = \frac{\sigma^2}{n}, \tag{4.2}$$

under conditions of sampling *with* replacement. In other words, this relationship holds when it is possible to draw an infinite number of samples or when the population is infinitely large. From our knowledge of expected value it is clear that an unbiased estimator of σ^2 will provide an unbiased estimator of $\sigma^2_{\bar{x}_n}$. This is truly an amazing result. We may provide an estimate (and a "good" one, perhaps) of the variance of the distribution of all n-sample means from only one sample!

Using our intuition again, a possible estimator of σ^2 is its analogue in the sample; i.e.,

$$\begin{aligned}\overset{*}{s}{}^2 &= \frac{1}{n}\sum_{j=1}^{n}(x_j - \bar{x}_n)^2 \\ &= \frac{1}{n}\sum_{j=1}^{n} x_j^2 - \bar{x}_n^2. \qquad (4.3)\end{aligned}$$

To evaluate (4.3) for the sampling with replacement case, we begin by taking its expected value:

$$\begin{aligned}E(\overset{*}{s}{}^2) &= \frac{1}{n}E\left(\sum_{i=1}^{n} x_j^2\right) - E(\bar{x}_n^2) \\ &= \frac{1}{n}\sum_{j=1}^{n} E(x_j^2) - E(\bar{x}_n^2). \qquad (4.4)\end{aligned}$$

Since we assume a sampling procedure where all random variables are independent and identically distributed, from the definition of the population variance

$$E(x_j^2) = \sigma^2 + \mu^2, \qquad (4.5)$$

so that

$$E(\overset{*}{s}{}^2) = \sigma^2 + \mu^2 - E(\bar{x}_n^2). \qquad (4.6)$$

By definition

$$\begin{aligned}\sigma_{\bar{x}_n}^2 &= E(\bar{x}_\mu - \mu)^2 \\ &= E(\bar{x}_n^2) - \mu^2. \qquad (4.7)\end{aligned}$$

Substituting for $E(\bar{x}_n^2)$ in (4.6),

$$E(\overset{*}{s}{}^2) = \sigma^2 - \sigma_{\bar{x}_n}^2. \qquad (4.8)$$

Under sampling with replacement $\sigma_{\bar{x}_n}^2 = \sigma^2/n$, which leads to

$$\begin{aligned}E(\overset{*}{s}{}^2) &= \sigma^2 - \frac{\sigma^2}{n} \\ &= \frac{n-1}{n}\sigma^2, \qquad (4.9)\end{aligned}$$

and *not* σ^2. Thus the sample value $\overset{*}{s}{}^2$ is *not* an unbiased estimator of σ^2 when sampling is done with replacement. But (4.9) suggests that

$$s^2 = \frac{n}{n-1}\overset{*}{s}{}^2 = \frac{1}{n-1}\sum_{j=1}^{n}(x_j - \bar{x}_n)^2 \tag{4.10}$$

is an unbiased estimate of σ^2, so that an unbiased estimate of $\sigma^2_{\bar{x}_n}$ is available from

$$s^2_{\bar{x}_n} = \frac{s^2}{n}. \tag{4.11}$$

The sampling without replacement case follows similar lines, but at step (4.9) it is *not* true that $\sigma^2_{\bar{x}_n} = \sigma^2/n$. Instead the proper form is†

$$\sigma^2_{\bar{x}_n} = \left(\frac{N-n}{N-1}\right)\frac{\sigma^2}{n}. \tag{4.12}$$

Inserting (4.12) in (4.8), we find

$$E(\overset{*}{s}{}^2) = \left[1 - \frac{N-n}{(N-1)n}\right]\sigma^2$$

$$= \frac{N}{N-1}\left(\frac{n-1}{n}\right)\sigma^2; \tag{4.13}$$

i.e., $\overset{*}{s}{}^2$ under sampling without replacement is again a biased estimate of σ^2 Moreover, using s^2 instead of $\overset{*}{s}{}^2$,

$$E(s^2) = \left(\frac{N}{N-1}\right)\sigma^2 \tag{4.14}$$

and not σ^2. But this difficulty may easily be overcome by defining a new population constant (numerically indistinguishable from σ^2 if N is large):

$$S^2 = \left(\frac{N}{N-1}\right)\sigma^2. \tag{4.15}$$

Then s^2 is an unbiased estimate of S^2, and we may rewrite (4.12) in terms of S^2 to get

$$\sigma^2_{\bar{x}_n} = \left(\frac{N-n}{N}\right)\frac{S^2}{n} \tag{4.16}$$

† We state this relationship here without proof. Cf. Cochran, *op. cit.*, pp. 22–23.

under sampling without replacement. Another common form for (4.16) is

$$\sigma_{\bar{x}_n}^2 = \left(1 - \frac{n}{N}\right)\frac{S^2}{n}. \tag{4.17}$$

The multiplier $[1 - (n/N)]$ is called the *finite population correction*, and is equal to the fraction of the population *not* sampled.

To obtain an unbiased estimate of $\sigma_{\bar{x}_n}^2$ under sampling without replacement, therefore, we use

$$s_{\bar{x}_n}^2 = \left(1 - \frac{n}{N}\right)\frac{s^2}{n}. \tag{4.18}$$

If n/N is small, (4.18) will closely approximate (4.11) numerically. Equation (4.18) *becomes* (4.11) as $N \to \infty$.

As an illustration of such calculations in practice, suppose a sample of 5 individuals from a population consisting of students in a large statistics class were drawn using sampling with replacement, and their heights measured. If the observed heights were 63″, 72″, 78″, 68″, and 69″, the relevant sample statistics would be $\bar{x}_5 = 70''$, $s^2 = 30.5$ in.^{2}, $s_{\bar{x}_5}^2 = 6.1$ in.^{2}, and $s_{\bar{x}_5} = 2.47''$. One way of presenting the results would be to quote the estimated mean as 70″ with a standard ("average") deviation of 2.47″.

Before turning to the question of effects on a normal confidence interval by using an estimated variance, consider the implementation of a normal approximation to the binomial or hypergeometric problems of estimating a population proportion. As we shall see in Chapter 6, our methods of inference are easily applied to these cases, using the exact distributions. But from a practical point of view it may be appropriate and much easier to use the normal approximation to these discrete distributions, which requires the expression of the estimated variance of the sample proportion, σ_p^2.

To give a one-to-one correspondence to our previous development, let the measured quantity X_i be defined over the population of N elements by the familiar zero-1 variable. Then the population proportion is $\tilde{P}$ and the population variance is given by

$$\begin{aligned}\sigma^2 &= \frac{1}{N}\sum_{i=1}^{N}(X_i - \tilde{P})^2 \\ &= \frac{1}{N}\sum_{i=1}^{N}X_i^2 - \tilde{P}^2 \\ &= \tilde{P} - \tilde{P}^2 \qquad \text{since } \sum_{i=1}^{N}X_i = \sum_{i=1}^{N}X_i^2 \\ &= \tilde{P}\tilde{Q}.\end{aligned} \tag{4.19}$$

Thus the (population) variance of the sample proportion, p_n, under sampling with replacement (binomial) is as in (4.2), with $\tilde{P}\tilde{Q}$ substituted for σ^2:

$$\sigma^2_{p_n} = \frac{\tilde{P}\tilde{Q}}{n}. \tag{4.20}$$

The analogous form for sampling without replacement is from (4.12):

$$\sigma^2_{p_n} = \left(\frac{N-n}{N-1}\right)\frac{\tilde{P}\tilde{Q}}{n}. \tag{4.21}$$

Following the development of (4.11), when sampling is accomplished with replacement and the population variance $\tilde{P}\tilde{Q}$ is estimated with the sample variance s^2, we see that

$$\begin{aligned} s^2 &= \frac{1}{n-1}\sum_{j=1}^{n}(x_j - \bar{x}_n)^2 \\ &= \frac{1}{n-1}\left(\sum_{j=1}^{n}x_j^2 - n\bar{x}_n^2\right) \\ &= \frac{1}{n-1}(np_n - np_n^2) \\ &= \frac{n}{n-1}p_nq_n, \end{aligned} \tag{4.22}$$

which provides an unbiased estimate of $\tilde{P}\tilde{Q}$ in (4.20) and results in the estimation equation

$$s^2_{p_n} = \frac{p_nq_n}{n-1}. \tag{4.23}$$

We know from our previous work that s^2 is also an unbiased estimate of $[N/(N-1)]\sigma^2$ for sampling without replacement. Substituting (4.22) in (4.21) for $[N/(N-1)]\,\sigma^2$ we get

$$s^2_{p_n} = \left(1 - \frac{n}{N}\right)\frac{p_nq_n}{n-1}, \tag{4.24}$$

when sampling is without replacement (hypergeometric).

As an illustration of the use of the normal approximation to a binomial sampling situation, suppose $n = 10$ and that 5 of the selected individuals possess the noteworthy characteristic. Then $p_{10} = .5$ and by (4.23) $s^2_{p_{10}} = .25/9 = .028$, $s_{p_{10}} = .167$, and *if* the normal is a good approximation to the binomial the 95% confidence interval for $\tilde{P}$ is approximately

$$P[.5 - (1.96)(.167) \leq \tilde{P} \leq .5 + (1.96)(.167)] \simeq .95\dagger$$

or

$$P(.173 \leq \tilde{P} \leq .827) \simeq .95, \qquad (4.25)$$

which states that this interval contains $\tilde{P}$ with probability .95. We shall return to this example in Chapter 6 for purposes of evaluating how good the normal approximation is to the actual binomial confidence interval.

This normal interval may likewise be used to produce confidence statements about a Poisson mean, λ. If in a sample interval of 24 hours the local post office received three special delivery letters, what is a 95% confidence interval around λ, the average number of special delivery letters received every 24 hours? Again, as in the binomial case, we are not yet prepared to give an answer to this question in terms of the exact distribution. But we may use the normal approximation with an estimated variance.

Recalling that the Poisson has both its mean *and* variance equal to λ, we may easily produce the required approximation:

$$P[3 - (1.96)(\sqrt{3}) \leq \lambda \leq 3 + (1.96)(\sqrt{3})] \simeq .95. \qquad (4.26)$$

In this particular instance, the lower interval bound ($-.39$) has little meaning. λ cannot, by definition, be either negative or zero. Thus a more precise statement would read

$$P(0 < \lambda \leq 6.39) \simeq .95. \qquad (4.27)$$

The basic modification in our formulas required to handle the Poisson case for a sample of n observations (rather than one, as above) is to note that for the Poisson, $\sigma^2 = \lambda$, so for the distribution of sample estimates $\hat{\lambda}$ we have, from (2.12):

$$\sigma^2_{\hat{\lambda}} = \frac{\lambda}{n} \qquad (4.28)$$

To estimate $\sigma^2_{\hat{\lambda}}$ is straightforward: simply substitute the sample estimate $\hat{\lambda}$ for λ in (4.28). Since it is an unbiased estimate of λ, the resulting estimated variance, say $s^2_{\hat{\lambda}} = \hat{\lambda}/n$, is an unbiased estimate of $\sigma^2_{\hat{\lambda}}$.

† With continuity corrections ignored.

One final word of caution: the normal distribution is in general a satisfactory approximation to the Poisson (cf. Section 5, Chapter 3) only for *very* large n, which is clearly *not* the case in the above illustration. We would expect this interval to be in error by a substantial amount—more than the indicated 5%. We use the example only to illustrate the mechanics of the application.

5. The t distribution

If the sampling distribution of n-sample means is *known* to be normally distributed (e.g., when the population distribution is normal), but the variance of the sampling distribution is *not* known, and is estimated by $s_{\bar{x}_n}^2$, then the statistic

$$t = \frac{\bar{x}_n - \mu}{s_{\bar{x}_n}} \tag{5.1}$$

does not follow the standard normal distribution. This is to say that the probability statement

$$P(\bar{x}_n - 2s_{\bar{x}_n} \leq \mu \leq \bar{x}_n + 2s_{\bar{x}_n}) \tag{5.2}$$

holds with probability *less* than .9544 as under the normal distribution. Or, saying it another way, at a probability .9544 and using $s_{\bar{x}_n}$ instead of $\sigma_{\bar{x}_n}$ the confidence limits $\bar{x}_n \pm 2s_{\bar{x}_n}$ are *too narrow*.

The basis for such discussion lies in three theorems about probability distributions, the first which we have already proved informally:

Theorem 1. *If y_i $(i = 1, \ldots, n)$ are normally and independently distributed random variables with means $E(y_i) = \theta_i$ and variances $E(y_i - \theta_i)^2 = \sigma_i^2$, then the weighted sum $\sum_{i=1}^n w_i y_i$, where not all $w_i = 0$, is normally distributed with mean $\sum_{i=1}^n w_i \theta_i$ and variance $\sum_{i=1}^n w_i^2 \sigma_i^2$.*

Theorem 2. *If z_j $(j = 1, \ldots, m)$ are independent standard normal random variables, then $\sum_{j=1}^m z_j^2$ is distributed according to Chi-squared (χ^2) with m degrees of freedom; $\sum_{j=1}^m z_j^2 \sim \chi_m^2$.*

Theorem 3. *If Z is N (0, 1) and x is χ_m^2 and independent of Z, then*

$$t = \frac{Z}{\sqrt{x/m}}$$

has the "t" distribution with m degrees of freedom.

We shall not attempt to prove either of the last two theorems, nor shall we give the mathematical forms of the χ^2 or t distributions. In particular, Theorem 2 is included to provide the necessary prerequisites to demonstrate that (5.1) possesses a t distribution, which like the standard normal is symmetric around zero.

The notion of *degrees of freedom* is important in theoretical statistics and, as stated in Theorem 3, is a parameter of the t distribution. The key to the term's meaning lies in Theorem 2: Degrees of freedom is the number of independent normally distributed variables in the sum $\sum_{j=1}^{m} z_j^2$. Interpreted for our purposes then, degrees of freedom relates to the number of observations used in a given calculation which are *unrestricted.* For example, in computing $\bar{x}_n = (1/n) \sum_{j=1}^{n} x_j$ there are no "restrictions" placed on the x_j's; they are indeed independent random variables and we have not altered that fact by combining them as $\bar{x}_n$.

Consider, then, the random variables $(x_j - \bar{x}_n)$, with $\bar{x}_n$ defined as usual $(j = 1, \ldots, n)$. Only $(n-1)$ of these variables are independent, since the "last" one must be such to meet the restriction $\sum_{j=1}^{n} (x_j - \bar{x}_n) = 0$. So, for example, the number of independent variables in

$$s^2 = \frac{1}{n-1} \sum_{j=1}^{n} (x_j - \bar{x}_n)^2 \tag{5.3}$$

is $(n-1)$. Another useful interpretation of degrees of freedom in this context is as the number of observations less the number of parameters estimated prior to any subsequent computation. For computing s^2 we need to have already calculated $\bar{x}_n$, the sample mean. Thus the number of degrees of freedom in s^2 is $(n-1)$.

We are now in position to utilize the theorems given to discover the distribution of t as given in equation (5.1). Given any normally distributed random variable, say, $x \sim N(\mu, \sigma_x^2)$, it can easily be transformed into standard normal form as

$$Z = \frac{x - \mu}{\sigma_x}. \tag{5.4}$$

Now if the population in question possesses a normal distribution, by Theorem 1

$$\bar{x}_n \sim N(\mu, \sigma_{\bar{x}_n}^2), \tag{5.5}$$

so that

$$Z = \frac{\bar{x}_n - \mu}{\sigma_{\bar{x}_n}} \sim N(0, 1). \tag{5.6}$$

Under sampling with replacement,† $\sigma^2_{\bar{x}_n} = \sigma^2/n$ and we use s^2 as an unbiased estimate of σ^2. We may use Theorem 2 to ascertain the distribution of $\sum(x_j - \bar{x}_n)^2$ only if each $(x_j - \bar{x}_n)$ has first been transformed to a $N(0, 1)$ basis. For the $(x_j - \bar{x}_n)$, $\bar{x}_n$ is a constant in the sample such that the variance of each $(x_j - \bar{x}_n)$ *given* the sample $\bar{x}_n$ is σ^2, the population variance.‡ Viewed now as a conditional variance (conditional on $\bar{x}_n$), we deduce that the variance of $(x_j - \bar{x}_n)$ is σ^2 no matter what $\bar{x}_n$ *is actually chosen*. Thus the conditional variance is also unconditional (i.e., independent of $\bar{x}_n$). The expected value of $(x_j - \bar{x}_n)$ is zero by the following reasoning: Given $\bar{x}_n$, $E(x_j - \bar{x}_n) = \mu - \bar{x}_n$. Now taken over all possible samples of size n, $E(\mu - \bar{x}_n) = \mu - \mu = 0$. Thus to correspond to the requirements of Theorem 2, we must deal in terms of the transformed variable $(x_j - \bar{x}_n)/\sigma$.

Then by Theorem 2,

$$\frac{\sum_{j=1}^{n}(x_j - \bar{x}_n)^2}{\sigma^2} = \frac{s^2(n-1)}{\sigma^2} \sim \chi^2_{n-1}. \tag{5.7}$$

Using Z of (5.6) in conjunction with (5.7) and invoking Theorem 3, we have

$$\frac{Z}{\sqrt{\sum_{j=1}^{n}(x_j - \bar{x}_n)^2/\sigma^2(n-1)}} \sim t_{n-1}$$

or

$$\frac{(\bar{x}_n - \mu)/\sigma_{\bar{x}_n}}{\sqrt{s^2(n-1)/\sigma^2(n-1)}} \sim t_{n-1} \tag{5.8}$$

from (5.7). Now using $\sigma_{\bar{x}_n} = \sigma/\sqrt{n}$,

$$\frac{(\bar{x}_n - \mu)/(\sigma/\sqrt{n})}{s/\sigma} \sim t_{n-1},$$

and, finally,

$$\frac{\bar{x}_n - \mu}{s/\sqrt{n}} \sim t_{n-1}. \tag{5.9}$$

† Or, sampling without replacement from this infinite population.

‡ The variance of a constant is surely zero. Thus the variance of $(x_j - \bar{x}_n)$ with $\bar{x}_n$ fixed is $\sigma^2_{x_j} = \sigma^2$.

A general confidence interval for μ, analogous to (4.1) but under the assumption that $\sigma_{\bar{x}_n}$ is estimated, is therefore available, and takes the form

$$P(\bar{x}_n - t_{\alpha/2,n-1}s_{\bar{x}_n} \leq \mu \leq \bar{x}_n + t_{\alpha/2,n-1}s_{\bar{x}_n}) = 1 - \alpha. \tag{5.10}$$

Using the data on heights of Section 4, where $\bar{x}_5 = 70''$, $s_{\bar{x}_5} = 2.47''$, a 95% confidence interval for μ is derived from cutoff points $t_{\alpha/2,4} = 2.77$, from Appendix Table 3. For the specific case,

$$P(63.16 \leq \mu \leq 76.84) = .95. \tag{5.11}$$

This interval is indeed wider than the analogous normal interval (with $Z_{\alpha/2} = 1.96$), or, conversely, for a given interval obtains a smaller probability. An important observation to note is that the t distribution becomes essentially described by the normal law with d.f. > 120. For most calculations, however, the inherent significance of the figures used will not be seriously disturbed by use of the normal whenever d.f. > 30.†

To reiterate, the t distribution applies *only* when the underlying distribution is normal. If, for example, we were using the (approximate) normal confidence interval for a population proportion even with n (or degrees of freedom) small we would *not* use t.

Finally, from (5.10) it is clear that if the underlying distribution of $\bar{x}_n$ is normal and we give a "point" estimate of μ with an accompanying estimated standard deviation as an indicator of the precision of our estimate, this presentation implicitly connotes a confidence level corresponding to $\bar{x}_n \pm 1s_{\bar{x}_n}$. For the example used above, the estimate of μ by $70'' \pm 2.47''$ implies (from a confidence interval based on the t distribution) a confidence level of approximately 60%, or 68% under normal assumptions.

† "Seriously" to again be interpreted in Chapter 7.

The hypergeometric distribution

Consider the possibility of obtaining two successes from a sample of $n=3$ under sampling without replacement. Obviously this outcome can occur in only three ways, observing the matter of order of occurrence in selecting the sample: (1) success on draw one, success on draw two, no success on draw three; (2) success, failure, success; and (3) failure, success, success. For configuration (1) the probability is determined thusly: Let A denote the total number of elements possessing the attribute in the population of N (so $\tilde{P}=A/N$). Then if at each stage in sampling all *remaining* items in the population are given equal chances of being selected, this configuration has a probability

$$\frac{A}{N}\frac{(A-1)[(N-2)-(A-2)]}{(N-1)(N-2)}=\frac{A(A-1)(N-A)}{N(N-1)(N-2)} \tag{5A.1}$$

of occurring, as opposed to

$$\frac{A^2(N-A)}{N^3}=\tilde{P}^2\tilde{Q}$$

when sampling with replacement. For the two remaining configurations we have

$$\frac{A}{N}\frac{[(N-1)-(A-1)]}{N-1}\frac{A-1}{N-2}=\frac{A(A-1)(N-A)}{N(N-1)(N-2)}, \tag{5A.2}$$

$$\frac{N-A}{N}\frac{A}{N-1}\frac{A-1}{N-2}=\frac{A(A-1)(N-A)}{N(N-1)(N-2)}, \tag{5A.3}$$

so that, as under binomial sampling, each configuration has the same probability of occurring, but the "total" probability of observing two successes is now

$$P(x=2\,|\,N, A, n=3)=\frac{3A(A-1)(N-A)}{N(N-1)(N-2)},$$

rather than $3\tilde{P}^2\tilde{Q}$.

This result is a simplification of the general from of the *hypergeometric* distribution, which may be written

$$P(x=k\,|\,N, A, n)=\binom{A}{k}\binom{N-A}{n-k}\bigg/\binom{N}{n}$$

The reader may wish to check that the general form will, in fact, produce our derived result by inserting $k=2$ and $n=3$. Also note that the denominator $1\bigg/\binom{N}{n}$ is just the probability that we shall obtain any one of the possible samples of size n from a population of N elements when sampling is accomplished with an EPSEM without replacement.

Comprehensive tables of the hypergeometric are available in Lieberman and Owen, *Tables of the Hypergeometric Distribution* (Stanford, 1961).

6

Classical hypothesis testing

A *statistical hypothesis* is merely a hypothesis about some population parameter or parameter set which influences the behavior of an observed random variable (a statistic). Thus if the statistic in question is an n-sample mean and the population mean is a parameter of interest, its value determines, through the frequency function $f(\bar{x}_n)$, the probability behavior of $\bar{x}_n$ either explicitly or implicitly.

The material in Chapter 5 makes reference to the probability that the observed difference between $\bar{x}_n$ and μ (in absolute-value terms) is no larger than some arbitrarily small number, or, as in the case of confidence intervals, the probability that certain (variable) limits contain μ. In this chapter we shall discuss statistical inference also, but in terms of formal tests of hypotheses about population parameters. As will become apparent, the two methods differ only in their orientation to problem solving and not in substance.

1. Statistical hypotheses and confidence intervals

Suppose we had taken a sample and computed interval limits for a given confidence probability. For a concrete example, let us use the sample of five heights from before, where the 95% confidence interval for μ based on the t distribution was

$$P(63.16 \leq \mu \leq 76.84) = .95. \tag{1.1}$$

The probability that the interval (63.16, 76.84) contains μ is .95. Of course, μ *could* be just about anything. That is, our sample of five observations *could* have been generated from a distribution with $\mu = 60''$ or $\mu = 80''$, etc. If μ is either of these values, the above interval obviously does not contain them. In fact, any values of $\mu < 63.16$ or $\mu > 76.84$ are *inconsistent* with the above interval where inconsistency has a probabilistic definition: *If we judge* the values $\mu < 63.16$ and/or $\mu > 76.84$ to be inconsistent with the observed sample interval at a 95% level of confidence, then the probability that this judgment is incorrect is 5%. Thus to reject any hypothetical mean in these ranges as inconsistent with the observed sample information implies a concomitant acceptance of a risk of error of rejecting such hypothetical values *when they are true* (called a type I error) of .05.

To elaborate, let us turn the argument around slightly and consider the probability distribution of 5-sample means from normally distributed population heights where $\mu = 78''$. This distribution is shown in Figure 6-1.

The analogue of a 95% confidence statement about μ in this situation

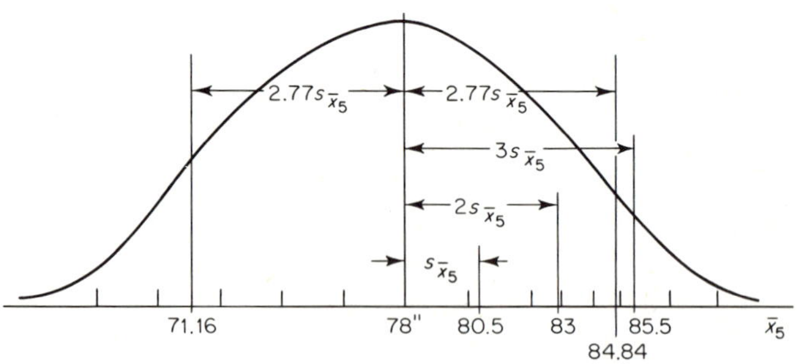

Figure 6-1. The t distribution of 5-sample mean heights with $\mu = 78''$, $s_{\bar{x}_5} = 2.47''$.

is simply a probability statement concerning the values of 5-sample means distributed around $\mu = 78''$; i.e.,

$$P(\mu - 2.77s_{\bar{x}_5} \leq \bar{x}_5 \leq \mu + 2.77s_{\bar{x}_5}) = .95 \tag{1.2}$$

or, substituting the proper values for μ and $s_{\bar{x}_5}$,

$$P(71.16 \leq \bar{x}_5 \leq 84.84) = .95. \tag{1.3}$$

In other words, 95% of 5-sample means from such a population would fall between these limits.

Suppose we decide to *reject* $\mu = 78''$ as inconsistent with an observed sample result whenever a 5-sample mean falls outside the above limits. Then clearly the probability that our rejection decision is incorrect *given* that $\mu = 78''$ is .05—simply the probability that sample means falling in this *rejection region* are observed under the hypothesis $\mu = 78''$. Using this decision rule with the example result $\bar{x}_5 = 70''$ we would reject $\mu = 78''$ as a tenable hypothesis, realizing that μ *could* be 78″ and accepting the probability of .05 being wrong in this respect.

Considering now replicating the above procedure for other possible hypotheses of the simple form $\mu = \mu_0$. Forming similar 95% probability statements about the limits to contain 5-sample means generated from populations with μ taking on various values and $s_{\bar{x}_5} = 2.47''$, we would find that the observed sample result of $\bar{x}_5 = 70''$ causes rejection (i.e., lies in the rejection regions) for all hypothetical population means $\mu < 63.16$ or $\mu >$ 76.84. The important implication is that a simple hypothesis about μ ($\mu = \mu_0$) may be tested at a given confidence level or *level of significance* (α) merely by constructing the confidence interval for μ based on the sample information and checking to see if the hypothetical μ_0 is contained within the interval. If it is, then at the given confidence level (set level of type I error probability) μ_0 should not be rejected as being inconsistent with the sample result. If μ_0 lies outside the interval it should conversely be rejected.

Here, then, is the bridge between statistical inference via an estimate (and some accompanying statement about its precision) and statistical inference in the form of a formal "test" of a hypothetical value of a parameter.

2. Confidence intervals for discrete sampling distributions

Confidence intervals for the discrete binomial and Poisson distributions may now be interpreted and constructed from exactly the above procedure of (1) considering the distribution under a particular parameter value; (2) finding the corresponding $(1 - \alpha)$ limits; (3) observing whether or not the sample result falls between the limits. The process is continued until the confidence limits are constructed. Of special note, however, is the fact that since these are discrete distributions one may not, in general, arbitrarily set α as we did for the continuous normal and t distributions.

To illustrate, consider the example of Chapter 5, where the observed sample proportion was $p_{10} = 5/10 = .5$. If we hypothesize that $\tilde{P} = .2$, set $\alpha \leq .05$, and attempt to split the total allowable risk (.05) equally into the two tails of the distribution (allowing no more than .025 in each) the resulting probability statement is

$$P(0 \leq p_{10} \leq .50) = .9936, \tag{2.1}$$

which indeed does meet the $\alpha \leq .05$ condition but not at equality. (Changing the upper boundary to $p_{10} = \frac{4}{10}$ would have increased $\alpha/2$ to .0328.) For the observed result $p_{10} = .5$ we would not reject $\tilde{P} = .2$.

Proceeding in a similar fashion, we find that the smallest $\tilde{P}$ for which our sample result would *not* have caused rejection at $\alpha \leq .05$ as above is $\tilde{P} = .19$ and the largest is $\tilde{P} = .81$ (roughly interpolating the Appendix table). Thus the "at least" 95% confidence interval for $\tilde{P}$ based on a sample of 10 and an observed sample proportion of 0.5 is

$$P(.19 \leq \tilde{P} \leq .81) \geq .95. \tag{2.2}$$

These exact limits compare very favorably with the interval (.173, .827) which was available from the normal approximation to the binomial interval at *exactly* a confidence probability of .95 (cf. Section 4, Chapter 5). And fortunately for users of statistics, such exact confidence intervals for the mean of both the binomial and Poisson cases have been constructed ad nauseum, so that computations such as those accomplished above need not be repeated nor use made of an approximate interval in most cases.†

3. Decision rules and errors in decisions

We shall call a fixed rule of behavior such as that used in the previous sections a *decision rule*. Such decision rules, then, cause the rejection of a hypothesis about some population parameter when the observed sample random variable falls in certain regions in its range, called rejection or *critical* regions, which control the probability of commiting an error of the sort: Reject $\mu = \mu_0$ when true.

† For the binomial, "Confidence Limits for p in Binomial Sampling," Tables 41 and 42 (pp. 204–205), and for the Poisson, "Confidence Limits for the Expectation of a Poisson Variable," Table 40 (p. 203) in *Biometrika Tables for Statisticians*, Vol. I, 2nd ed., edited by E. S. Pearson and H. O. Hartley (Cambridge, 1958).

The risk of rejecting a true hypothesis about a parameter like μ is only half the story, however. Consider ex post that the sample statistic for testing a hypothesis about μ did not fall in the rejection region at a given α-level. Our adopted decision rule says not to reject such a hypothesis. But as it was *possible* for any particular sample result to be generated for the hypothesized value of μ, so is it possible that a sample result which does not dictate rejection comes from a population with some other governing parameter. In brief, another type of decision error (type II) derives from *not* rejecting a hypothetical parameter when it is *not* correct.

To formalize these ideas, let us represent a hypothesis about some parameter in a population as H. We shall also limit ourselves to discussing hypothesis tests about a mean μ, although the principles involved may easily be extended to tests about any population parameter or a set of parameters tested simultaneously. For instance, a hypothesis about the mean height from before is H: $\mu = 78''$. Since the parameter is hypothesized to have a single value, 78″, this is called a *simple* hypothesis. Any hypothesis which contains more than one simple hypothesis is a collection of them and termed *composite*. H: $\mu > 78''$ is an example of a composite hypothesis about μ.

Suppose H: $\mu = 78''$ is a relevant hypothesis to test, and that as an estimator of μ we use, as before, $\bar{x}_n$. Given that observations come from a normally distributed population in terms of height, then $\bar{x}_n$ theoretically follows a normal distribution also if $\sigma^2_{\bar{x}_n}$ (from σ^2) is known. Since in the example we used before $\sigma^2_{\bar{x}_n}$ was estimated (as will be the usual case in actuality), we know therefore that the t distribution governs the distribution of n-sample means around μ, although empirically we may revert to the normal as an approximating distribution if n is large enough. Finally, it is known that μ is the expected value of $\bar{x}_n$, so that the variance of $\bar{x}_n$ is the proper precision measure. With these preliminaries completed, we are ready to discuss the hypothesis testing problem from another angle, that of H: $\mu = 78''$ being false.

If H is false, then some other value of μ is obviously the true one; i.e., $\mu \neq 78''$. Thus if H is rejected, $\mu \neq 78''$ is the adopted or implied alternative hypothesis. There are many possible alternative hypotheses which might be used as the datum for a particular problem, but for now we shall look specifically at the general alternative hypothesis H_A: $\mu \neq 78''$, which corresponds to the analogous test from the constructed confidence interval for μ.

What values for $\bar{x}_n$ are inconsistent with H as compared to H_A? From before, the decision maker may easily control α, the probability of rejecting H when it is true. But what of the chances of "accepting" H when it is false? The confidence-interval approach to the problem described inconsistency at some α-level in terms of the sample mean deviating by more (absolutely) than some amount from μ. If H is false, however, the confidence interval has little to say concerning this second sort of error.

Let H be assumed false, i.e., μ has some value other than 78″. For any given single value of μ, say, $\mu = 76''$, we can compute the probability that $H: \mu = 78''$ will be rejected.† Choosing $\alpha = .05$ as before, the t rejection regions for a 5-sample mean are $\bar{x}_5 < 71.16$ and $\bar{x}_5 > 84.84$. Now if μ *really* is 76″, we can compute β, the probability that $\bar{x}_5$ falls in (71.16, 84.84), which by our fixed rule does not cause rejection of $H: \mu = 78''$ when (by assumption) $\mu = 76''$. Pictorially the required probability is between 71.16 and 84.84 on a t distribution with $s_{\bar{x}_5} = 2.47''$, $\mu = 76''$, and 4 d.f. Converting to standard deviation units and zero mean, these limits become

$$t_L = \frac{71.16 - 76}{2.47} = -1.96,$$

$$t_U = \frac{84.84 - 76}{2.47} = 3.57. \tag{3.1}$$

The probability that t will lie within these bounds is approximately .925, roughly interpolating Appendix Table 3 for 4 d.f. Thus, if $\mu = 76''$, the probability that our test (decision rule) will err in *not* rejecting $H: \mu = 78''$ is $\beta \simeq .925$.

Of course, this probability will clearly change as other *particular* values of μ in the alternative range are considered. As μ is shifted further left (from 76″) in Figure 6-2, *less* area will be contained within the rejection limits, and

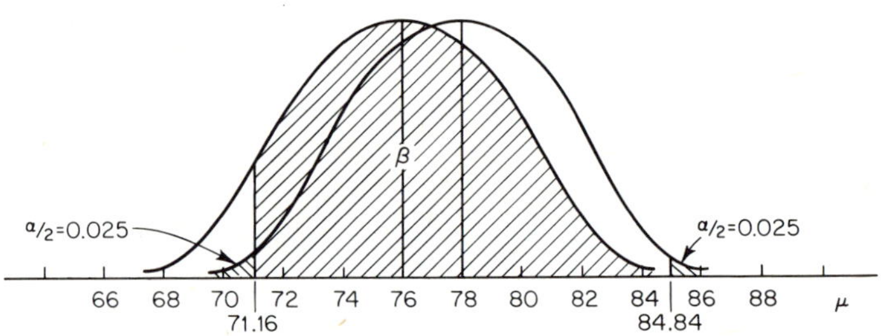

Figure 6-2. Computation of β, the probability of "accepting" a false hypothesis for the case $H: \mu = 78''$ vs. $H_A: \mu \neq 78''$ at $\mu = 76''$, $s_{\bar{x}5} = 2.47''$; t distribution, 4 d.f.

as μ comes closer to 78″ the opposite reaction of β is noted. As μ *becomes* 78″, the probability of nonrejection of a false hypothesis has no meaning, since $\mu = 78''$ *is* the main hypothesis being tested. The probability of not rejecting it when it is true is just $1 - \alpha$, the confidence probability.

Since the sampling distribution, t, is symmetric, like calculations for $\mu > 78''$ are mirror images of those just contemplated. The result of such calculations, covering *all* values $\mu \neq 78''$, may be summarized in a graph such

† Strictly speaking this computation requires the use of a "noncentral" t distribution. For our purposes we shall assume the "central" t is an adequate approximation (which it likely is not in this case), and proceed to view our results as suggestive rather than exact.

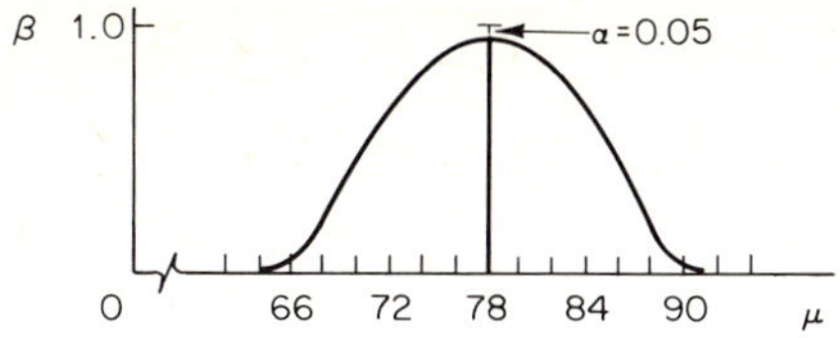

Figure 6-3. Graph of β, probability of "accepting" H: $\mu = 78''$ when it is false, $n = 5$, t distribution.

as Figure 6-3, where β is tabulated for the fixed decision rule: Reject H: $\mu = 78''$ in favor of H_A: $\mu \neq 78''$ whenever $\bar{x}_5 < 71.16''$ or $> 84.84''$.

The performance a priori of the fixed rule may be evaluated at least tentatively via an interpretation of this figure. Essentially the fixed rule—at the given sample size—performs well if the true μ is 8″ or so away from 78″. "Performs well" needs further definition, and it will be given in Chapter 7. But for now we shall take it to mean "small" β for the already arbitrarily small α.

Is it possible to design a rule for which $\beta = 0$ at some α-level and for a given sample size? From the figure, to obtain $\beta = 0$ requires that there be a zero probability of sample means in the "acceptance" region being produced from any other μ than $\mu = 78''$. Of course, this state of affairs is unattainable. But if we allow α to change (increase) we may influence β at a given sample size. Increasing α implies *less* confidence in any particular interval containing μ; or for hypothesis testing, increasing α implies narrower rejection limits to the extreme that when $\alpha = 1$ the rejection interval (for the case at hand) *is no longer* an interval. Thus to obtain $\beta = 0$ we may set $\alpha = 1$ and always reject H no matter what the sample result. Logically, of course, if we always reject H we most certainly will *never* be guilty of not rejecting it when H_A is true.

Conversely, we may design a rule for fixed n which will allow $\alpha = 0$. This rule, by a similar argument to that above, would dictate "acceptance" of H *always*. Thus we would definitely not err in rejecting H when true, but would always err in "accepting" it when H_A is true. How to best design a rule which takes into consideration the weights to be applied to making such errors—their consequences to the decision maker—is the climax to the present development and is treated in Chapter 7. However, there is another variable in the design procedure to consider first, which may be adjusted to reduce β probabilities over the range of α *without* altering α; that variable is sample size.

4. Effects of sample size on the probabilities of error in decisions

Consider exactly the same test as before under the changed condition that a sample of 25 height observations is to be taken rather than 5. And, instead of assuming a hypothetical set of values, computing an estimated population variance, and finally estimating $\sigma^2_{\bar{x}_{25}}$, let us also assume that the estimated population variance for these 25 heights turns out to be the same as for the five.† So we assume $s^2 = 30.5\text{in.}^2$ as the estimate of σ^2 in the population and thus $s/5 = 1.1''$ is then the estimate of $\sigma_{\bar{x}_{25}}$.

For purposes of generating the rejection regions for a test of H: $\mu = 78''$ versus H_A: $\mu \neq 78''$ at $\alpha = .05$ we shall also assume that the normal distribution yields a satisfactory approximation to the t distribution. The .05-level rejection points are, therefore, given as $78 \pm (1.1)(1.96)'' = 78 \pm 2.16''$.

To construct the β-function for this test we proceed as before: Let $\mu = 76''$. Then the rejection points in terms of this distribution occur at $(75.84 - 76)/1.1 = 0.15$ and $(80.16 - 76)/1.1 = 3.78$. The area contained under the standard normal in this interval is .4404, as compared to a β-probability of .925 with $n = 5$. Similar calculations for other values of μ yield Figure 6-4,

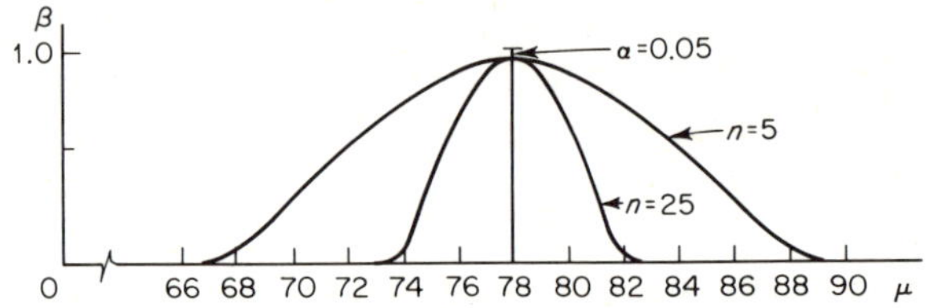

Figure 6-4. Graph of β, probability of "accepting" H: $\mu = 78''$ when it is false, $n = 25$, normal distribution.

and the resultant effect of an increase in sample size from $n = 5$ to $n = 25$ is clear as we superimpose the previous β-function on that for $n = 25$.

Such behavior is consistent with a previous observation concerning sample-size effects on confidence intervals about μ. For given confidence probability $1 - \alpha$ we noted that an increase in sample size accomplished a tightening of the confidence limits, which here corresponds to a decrease in β-probability over the entire range of the alternative hypothesis.

The decision maker's optimization problem, therefore, concerns discovering the "best" test of a given hypothesis vs. an alternative hypothesis about μ in the sense of balancing, according to some a priori system of weights, α and β probabilities in *conjunction* with sample-size determination, assuming it "costs" something to sample. And, although in this chapter we shall concern ourselves solely with questions of "bestness" given a sample size, the decision maker's choices in reality are broader than this limited analysis suggests, since a decision to sample or not *is* an alternative action in actuality.

5. Notion of a "best" test

Although it is intuitively appealing to reject H: $\mu = \mu_0$ vs. H_A: $\mu \neq \mu_0$ for extreme sample values, intuition alone cannot provide the objectivity

† This assumption is usually reasonable based on actual observation and computations. Although there is, most definitely, an inverse relationship between $\sigma^2_{\bar{x}_n}$ and n, there is no such a priori relationship between s^2 and n. Thus, although we should expect the population variance σ^2 to be estimated *more precisely* with larger n, the *actual* estimate may not be much different for large n than for small n.

required to make the methods of statistical inference generally usable and unambiguous. Moreover, as was demonstrated in Chapter 2, intuition may be a poor guide in questions dealing with random events and random variables.

To illustrate how intuition based on the actual problem situation may imply a particular rejection region for a test, suppose a garment maker wishes to discover whether the average height of American males has increased over its previous value of 10 years ago, estimated at 68″, for the purpose that if sufficient evidence can be produced to refute the hypothesis of status quo in favor of an increased average, the garment maker will add greater proportions of long suits to his output.

Intuitively, no matter what sample size might be used, a sample result will be inconsistent with the status quo hypothesis versus $H_A\colon \mu > 68''$ only if the sample mean is "too large." If $\bar{x}_n$ is observed at 65″, for example, it would be "unreasonable" to reject $H\colon \mu = 68''$ in favor of $H_A\colon \mu > 68''$. For many distributions of heights, e.g., the normal, this intuition may be reinforced logically from the development of a notion of what constitutes a desirable or "best" rejection region (test) for a given sample size. But using the adopted criterion, in general such logical support is not possible.

The classical criterion of "bestness" is simple: First, to reduce the number of variables in the problem, the α-level is kept constant in comparing alternative critical regions. Then the criterion is defined as follows: A critical region CR_1 is judged superior to an alternative CR_2 if for given α, $\beta_1(\mu) \le \beta_2(\mu)$; that is, if the β-function for CR_1 is always on or below the corresponding β-function for CR_2 over the range of H_A.

Use of the criterion implies that the testing problem concerns a simple hypothesis only, so that α is a single value and may, indeed, be held constant. If this is not the case the classical criterion is ineffective in determining which of the many possible alternative critical regions to use in a particular problem. This impasse forms the focal point for the modern contribution, and is treated in Chapter 7.

To illustrate how the classical criterion is utilized, let us suppose that a very good estimate of the variance of heights of American males is available, that heights are normally distributed, and that a sample of 100 persons is contemplated. The "reliable" value for σ^2 is 100, so that $\sigma^2_{\bar{x}_n} = 100/100 = 1$. In this case the FPC may certainly be neglected. Finally, we assume an α-level of .10.

The basic alternative critical regions for a test of $H\colon \mu = 68''$ versus $H_A\colon \mu > 68''$ are three: $\bar{x}_{100} \ge C$, $\bar{x}_{100} \le C$, or a combination of the two, $\bar{x}_{100} \le C_1$ or $\bar{x}_{100} \ge C_2$. We shall explicitly treat a comparison of the first two cases and argue the nonoptimality of the last from those results.

At $\alpha = .10$ the standard normal cutoff point is 1.28; thus the "intuitive" test suggests a critical region $\bar{x}_{100} \ge 68 + (1.28)(1)''$ or $\bar{x}_{100} \ge 69.28''$. The "opposite" test is $\bar{x}_{100} \le 66.72''$ and the two-tailed test is some double critical region with $C_1 > 66.72''$ and $C_2 < 69.28''$ (since α is set at .10).

By calculating the β-functions for the first two tests we can hopefully apply the classical criterion to determine which test is best. Figure 6-5 shows

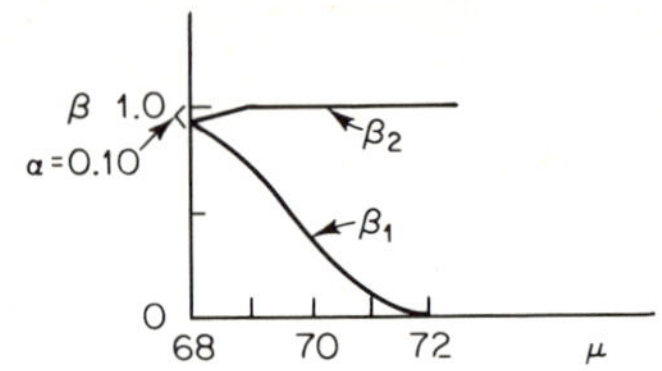

Figure 6-5.

the relevant functions, where it is clear that the critical region $\bar{x}_{100} \geq 69.28''$ is superior. For a fixed $\alpha = .10$, it provides smallest probability of "accepting" H when *any* μ in H_A is true.

If we construct a two-sided critical region, keeping $\alpha = .10$, it is also clear that $\bar{x}_{100} \geq 69.28''$ would remain superior. The argument is as follows: Beginning with the upper region only, we have β_1 of Figure 6-5 as the governing description of type II error probability behavior over μ in H_A. To include a lower region is tantamount to increasing the total β over all μ in H_A, since the lower region possesses an increasing β-function over μ. Any substitution of lower region for upper region, therefore, encompasses a combination of the β_1 and β_2 functions which will lie somewhere between them. For the "well-behaved" normal distributions, then, our intuitive critical region is also best in the classical sense.

As there are methods for producing estimators with certain properties there are techniques for generating tests which likewise possess desirable properties, such as "bestness" (as defined herein). Although we shall not discuss these methods at any length here, the reader should note that they too depend on the ability to control α, the probability of committing a type I error.

Example of incorrect intuition in designing a decision rule†

Consider a random variable x which has the distribution $f(x) = \theta e^{-\theta x}$. As long as $\theta > 0$ and $x \geq 0$, $f(x)$ is a valid frequency function, i.e.,

$$\int_0^\infty \theta e^{-\theta x}\, dx = -e^{-\theta x}\Big|_0^\infty = +e^0 = 1, \tag{6A.1}$$

and has a graph of the type shown in Figure 6A-1.

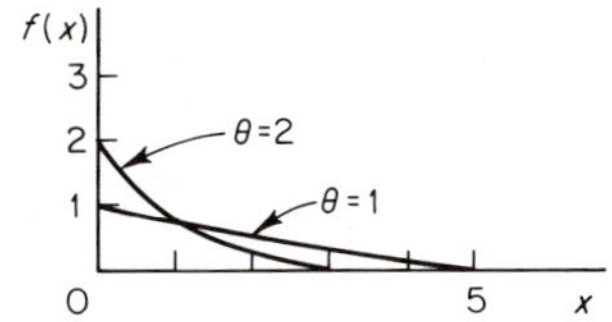

Figure 6A-1.

Suppose we wished to test the hypothesis $H: \theta = 2$ versus the alternative $H_A: \theta < 2$ for a sample of size 1. An "intuitive" rule for this test is: Reject H in favor of H_A if $x \leq c$. The comparison we shall attempt here is between a rule of this form and one with a critical region $x \geq 1$.

For the region $x \geq 1$, α-probability may be obtained from

$$\alpha_1 = \int_1^\infty 2e^{-2x}\, dx = .135, \tag{6A.2}$$

and β-probability through noting that we may write

$$\beta_1(\theta) = \int_0^1 \theta e^{-\theta x}\, dx = 1 - e^{-\theta}. \tag{6A.3}$$

The subscript used in α_1, etc., denotes reference to "test 1" with critical region $x \geq 1$.

A competing test $x \leq x_0$ of the "intuitive" form with the same α-probability (.135) may be obtained by requiring

$$\alpha_2 = \int_0^{x_0} 2e^{-2x}\, dx = .135, \tag{6A.4}$$

which leads to

$$\begin{aligned} 1 - e^{-2x_0} &= .135, \\ \log_e(.865) &= -2x_0, \\ x_0 &= .072. \end{aligned} \tag{6A.5}$$

† Adapted from P. G. Hoel, *Introduction to Mathematical Statistics*, 3rd ed. (Wiley, 1962), Chap. 3.

Thus at a constant α-level we may compare the ability of the tests to discriminate between H and values of θ in H_A when H is false.

For test 2 the β-function may be obtained simply by

$$\beta_2(\theta) = \int_{.072}^{\infty} \theta e^{-\theta x}\, dx = e^{-.072\theta}. \tag{6A.6}$$

The functions $\beta_1(\theta)$ and $\beta_2(\theta)$ are shown in Figure 6A-2 for θ in H_A, where it is

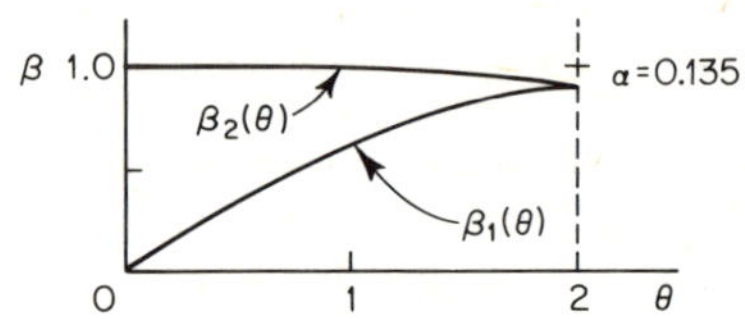

Figure 6A-2.

quite clearly demonstrated that the critical region $x \geq 1$ is superior to the "intuitive" region $x \leq .072$ at the given α-level, .135.

The problem here is not with the classical statistics, but with the "intuition." θ, in fact, is not the mean of the distribution; $1/\theta$ is. Therefore small values of x are consistent with the main hypothesis rather than the alternative, in exact opposition to the case as written. As it turns out, the correct "intuitive" test is also "best."

7

Costs of decision errors and the optimal decision rule

The classical approach to two-action testing problems, although providing a logical framework for decision making, cannot always give us a "best" rule for a particular problem in terms of error probabilities alone. Thus when such errors are evaluated also in terms of the cost of their consequences, it is not surprising that the classical method meets a similar impasse, whether the costs of errors be stated merely as relative or implicit, in monetary or other terms.

When the decision problem can be stated with a simple hypothesis and a simple alternative, obviously the classical method will produce a best rule, since we may equal α-probabilities (weighted or unweighted) and look for that rule which possesses smallest β (again, either weighted or unweighted). There may be more than one rule which provides minimum β under the conditions stated, but we shall be able to find an optimum decision rule to follow.

When the alternative hypothesis becomes composite, our ability to find an optimum rule is impaired by the fact that the criterion of bestness must be altered from what it was for the simple alternative case: The best rule now must provide minimum β for all values of the alternative hypothesis. Restated, the type II error function for the optimum rule must lie on or beneath all other such functions (corresponding to all other competing tests). Again, the matter of weighting α and β by costs to indicate their relative importance is straightforward but not helpful in solving this dilemma, since the weights will affect the error functions in a similar manner.

If we take the final step and look at the composite hypothesis, composite alternative case, we must consider varying consequences of a type I error for parameter values in the range of the main hypothesis. So if application under cost considerations is our primary goal, apparently the classical approach leaves something to be desired.

One might consider making some choice of a rule by value considerations. That is, faced with a graph of the type II error functions under the simple hypothesis–composite alternative case, for example, where, for the sake of argument, clearly no best rule exists under the classical criterion, one might reason: "Even though rule A is not best by classical standards over all values of the alternative, it is best over those values which I feel are most relevant—most likely to be the values of the *true* parameter in question, so I will use rule A." Wait a minute! One of the major premises of the classical framework is that there is a *parameter*—an unknown, yet *single-valued* function of the random variable in question. How then can we even consider discussing *values* of a parameter as if one value were more likely than another, etc., for example, giving the parameter a frequency interpretation? Does not such an interpretation void the classical framework?

These are the questions which we shall attempt to answer in this chapter. The most important results, in terms of impact on the methodology of decision making under uncertainty, which appear from an investigation of these questions derive from a change in the classical interpretation of a parameter. What evolves from such a change is an addition to our logical framework, not a destruction of what we have spent six chapters to present. In fact, the new interpretation provides a synthesis of classical methods and their relation to how decisions are really made. The net result is a simulation of the decision process and, necessarily, a basis for making normative statements about how decisions should be made.

1. The parameter as a random variable

Suppose we conceptualize that a parameter is not, as the classical interpretation would dictate, a *fixed* but unknown value in a defined population. Rather, the parameter itself is a random variable: It has a probability distribution, a mean, variance, etc.

Many arguments may be offered to support such a premise, among them the fact that even though a population may be "well defined," its basic structure may be subject to change such that at any instant of time the parameter in question has a fixed value, but in a dynamic context the parameter can actually take on a variety of values simple because the population is subject to change over time.

Another fruitful way of looking at the parameter concept is to consider problems of measurement. Even though the meaning of the term parameter may be adhered to in its classical interpretation conceptually, do measurement errors exist? Suppose we sampled an entire population and calculated some parameter. Is this value really the parameter? Are our techniques of measurement so precise as to preclude the possibility of error? Hardly. Even the most naïve interpretations of the characteristics of measurement error suffice to enforce the probabilistic interpretation of a parameter. Specifically, this parameter which exists may never be measured exactly, hence we should be concerned, as a practical matter, with the *observed* value of the parameter, which is a random variable if measurement error is considered random.

There are other, more sophisticated arguments surrounding the issue of interpretation which has been raised, both in favor and opposing the probabilistic view, which we shall not go into here. Most important, the student should note that this issue is *the significant difference* between the classical and "modern" schools of thought, and, as we shall see, a modern interpretation opens many doors to the resolution of previously "unsolvable" problems in statistical decision making.

Moreover, it is contended that decision makers do indeed *act as if* the parameters of their decision problems were variables, and make value statements such as that posed in the preceding section; hence the modern interpretation of a parameter is really a step toward the explanation of real-world decision making, with two definite ramifications: First, the modern interpretation of a parameter, if it holds as a premise for decision making under uncertainty, leads to the formulation of a theory of decision making. In turn this theoretical structure provides a logical framework which we may then use to make normative judgments about decision making.

As will be made clear in what follows, the modern interpretation does not destroy the classical theory which preceded it. On the contrary, with only a small amount of redefinition and some new terminology, the result is an almost complete absorption of classical methodology into a new logical scheme.

As with many conceptual "revelations," however, new problems arise concomitantly with a modern interpretation. These problems are primarily of application, and are concerned with the "value" of information, type of information, etc., about the parameter in question. Succinctly, how does one go about stating the distribution of the parameter from available information? And, what if no a priori information is available, then what? These, among other questions, are most critical in practice. They will be posed again in Section 9, and engaged briefly there.

2. The setting—establishing costs of decision errors

We shall consider a simple problem in decision making for the purpose of developing general concepts in the modern approach. As with many applications of statistical methods, both classical and otherwise, one of the most difficult jobs is the translation of the real-world problem into terms by which a statistical model may be constructed to analyze the problem. Our approach in this book has been to build "models"; that is, by assumptions, simplifications of reality, etc., we have produced simulators of reality. When we solve a decision-making problem involving the normal distribution for a particular statistic we first assume the normal distribution. If you were to confront a real problem which corresponded to all the other conditions and assumptions made in our exercise, you still would need to assess the correspondence of your sampling distribution to the normal curve.

Perhaps some other distribution is appropriate on theoretical grounds, such as the binomial distribution for a coin-tossing problem. Or perhaps the distribution must be approximated by fitting a known curve to some sample data. In any case, the point to be made is that each time a problem is encountered a new "model" must be constructed. Very rarely do "textbook" examples occur.

When we previously spoke of weighting decision errors by the "costs" of making such errors, we passed over a difficult practical problem, because monetary costs usually do not describe a particular situation adequately and in some situations monetary costs are not directly involved at all. For example, the loss of 10 dollars to a millionaire would likely be of less consequence to him than the same loss to a pauper. The same sort of difference in the "value" of money probably explains why some people play poker for pennies and others play for dollars.

Since money obtains its value through the goods and services it represents (at least it used to), a given sum will have different "value" to different people, just as the possession of certain goods will have different value to the same people. The dilemma we must face involves those situations in which dollar amounts and their "values" do not bear a proportional relationship to one another, for clearly if a proportional relationship exists it will not matter whether we have monetary or "value" measures to assess the *relative* costs of type I and type II errors.

How do we discover that this proportional relationship does or does not exist over the relevant dollar range for the problem at hand; how do we construct the necessary value, or, as it is more commonly referred to, *utility* scale for money? Again, here is a prerequisite to a textbook application which may not be (indeed, most probably is not) met in a real-world problem. Hence, before the use of any method of analysis which treats the weighting of decision errors, we must produce another model to handle the money-utility inequality which may exist.

A suggested method for constructing the individual's utility scale for money in a given decision problem is available in Schlaifer's *Introduction to Statistics for Business Decisions*,† and as our method of analysis for the "best" decision rule is independent of the measure used to determine the cost of decision errors, the reader is referred there. In the material that follows, therefore, monetary costs will be assumed to hold as "real" costs. When necessary we shall annotate the development to include any pertinent analogies to the decision problem with a utility rather than monetary cost valuation.

3. An example‡

Suppose a prospective entrepeneur is considering embarking on a career as a pearl dealer. This fellow is very scientific in his outlook toward decision making, having taken a meaty course in statistics in college; however, he has requested that we assist him in the analysis. To this end he gives us the following information. The oyster bed in question contains 500 oysters (counted by scuba divers, each using an abacus adapted for underwater use). It would cost our operator-friend \$200 in fixed costs to set up his equipment; average variable costs amount to \$1 per oyster for gathering, opening, etc. The operator can sell pearls for an average of \$4 per pearl, but cannot sell the oysters themselves at all.

By doing some arithmetic we discover that if there are more than 35% of the oysters with pearls the operator will make a profit; at exactly 35% he breaks even, and if the proportion with pearls, say $\tilde{P}$, is less than .35, he loses money. In his estimation the time he will spend digging up 500 oysters could be used selling newspapers, for which he would receive \$300. So, he is

† (McGraw-Hill, 1961), Chap. 2.

‡ We shall take this opportunity to present the material to follow as a "case" of application as well as a vehicle for introducing the modern contribution. Thus many restatements of previous topics appear, to point out how the "synthesis" is accomplished in an actual application, particularly in Section 4.

primarily interested in ascertaining whether or not his endeavors in the oyster game will reward him at least as well as selling newspapers. If the proportion of oysters with pearls is at least .50, he will make \$300 or more; hence this seems to be the relevant hypothesis to test.

We now face two problems: the determination of sample size and the choice of a decision rule for testing the hypothesis. To begin with we shall arbitrarily select a sample size of $n = 20$ and assume it costs nothing to obtain the sample. Later both these assumptions will be relaxed and included as part of our total analysis. As to the hypothesis to test, we have already indicated that the region $\tilde{P} \geq .50$ is relevant as against $\tilde{P} < .50$, where the operator definitely loses money, either as a direct loss ($\tilde{P} < .35$), or an opportunity loss ($.35 \leq \tilde{P} \leq .50$). At this point we shall further assume that our friend's utility scale for money is linear, so that monetary costs and profits do indeed describe his valuation of all losses and gains.†

Our task is to determine a decision rule for testing H: $\tilde{P} \geq .50$ against H_A: $\tilde{P} < .50$ for $n = 20$, which on the average provides smallest losses from making decision errors.‡ Clearly, the pertinent considerations for selecting the rule involve incorrect decisions and the losses resulting from them. For example, suppose $\tilde{P}$ is .50 but we reject H. Our costs from making this error (type I) will be zero, since the opportunity loss of not "digging" when it is profitable to do so is \$300, but we can sell newspapers for \$300. The other sort of incorrect decision we could make is not to reject H when, in fact, H_A is true (type II error). Again, as an example, our losses (direct) from this incorrect decision would amount to \$100 plus \$300 in opportunity losses from not selling newspapers, for a total of \$400 if $\tilde{P}$ were really .3. In fact, we can generate relationships which will describe such total losses as a function of $\tilde{P}$: losses $= \$2000\tilde{P} - \$1000$ if $\tilde{P} \geq .5$, and losses $= \$1000 - \$2000\tilde{P}$ if $\tilde{P} \leq .5$. Obviously such losses are precisely determined by the value which $\tilde{P}$ actually attains in the population of 500 oysters, and hence could be called "conditional" losses, conditional on $\tilde{P}$.

We shall use a concept analogous to the frequency definition of probability in choosing our best decision rule. For us a "trial" consists of taking a sample of 20 from the population of 500 and either rejecting or not rejecting the main hypothesis on the basis of this sample. What we require is a decision rule which would guarantee that if we repeated the same test, each time drawing a new sample of 20 and either rejecting or not according to our fixed rule, the losses from making incorrect decisions be at a minimum. As we shall see, to meet this criterion requires the modern interpretation of a parameter.

† If this function were not linear (proportional) in money all the calculations would be carried out in terms of utility measure.

‡ Since the equality $\tilde{P} = .50$ represents a breakeven point, it could be included in either the main or alternative hypothesis.

4. Determining the sampling distribution of the test statistic

How should the sample be selected? This question again poses the problem of adherence to assumptions, in the sense that: (1) the methodology of statistical testing requires a probability sampling method, and (2) in any real application the appropriate sampling design must be selected by cost considerations in relation to the required precision of results. When we discussed sample-size determination in dealing with classical testing problems, we only considered simple random sampling methods and merely indicated that somehow the costs of obtaining a given size sample should be considered in deciding what particular sampling method to use. There are reams of literature on the subject of sample design, and for our purposes here we must limit alternatives, so simple random sampling will be stressed again in this example. The important thing to observe is that no matter what probability sampling method is used, the inherent nature of the method will have different implications for the sampling distribution of a given test statistic. But they all will have one basic thing in common, that of assigning to each possible combination of n items from the population a definite probability of selection. The simple random methods, both sampling with and without replacement are EPSEM methods—equal probability selection methods. They assign equal weight (probability) to each possible sample.

What are the effects of each simple random method on the distribution of a given test statistic? To return to our specific case, since we are concerned with losses resulting from the proportion of oysters being small, it should be clear that our decision rule will be formulated to tell us when the observed sample proportion is so small as not to be consistent with the main hypothesis. That is, the rule will state that we should reject H whenever the observed number of oysters with pearls in the sample of 20, say y, is smaller than some number c, where c is some possible value of y.

The distribution of our test statistic, y, will depend on the method of sampling used. First, as a general proposition, we know our problem concerns a basic random variable which can only take on two "values," pearl or no pearl.† The selection of an oyster may be considered a trial, and the possible outcomes of each trial the events pearl and no pearl. Each outcome has a certain probability of occurring, the probabilities being determined by the proportions of oysters with and without pearls in the population; that is, $\tilde{P}$ and $\tilde{Q} = 1 - \tilde{P}$, from the definition of probability.

As long as the probability that any oyster contains a pearl remains constant as we take our sample, the resulting distribution of the total number of oysters with pearls in the sample will theoretically follow the binomial

† Note that we are assuming away measurement problems, so there are only two possible outcomes: We can always ascertain whether a pearl exists or not under the assumption.

distribution. But the requirement of $\tilde{P}$ remaining constant as we sample is only met when we sample with replacement, where on each "trial" (or draw) the population is unchanged; previous oysters are always replaced into the population before a new selection is made. Thus $\tilde{P}$ (in the sampling sense) will remain constant throughout the entire process of selecting 20 items from the 500. Moreover, the proportion $p_n = y/n$, the sample proportion of oysters with pearls, will also follow the binomial distribution, since y and p_n differ only by a constant factor of proportionality, n. That is to say, if the probability that y has the value 10 is .1762 for a sample of 20, then the probability that p_{20} is .5 (10/20) is also .1762.

When sampling is done without replacement, the situation is slightly altered by the fact that at each stage of the sampling procedure the probability that the oyster selected has a pearl is *dependent* upon the results of previous selections. For the example, once an oyster is opened, its physical nature is somewhat "destroyed"; it really is not the same object as the oyster which remain unopened. Thus to sample with replacement would require some change in definition for the population. Similar, but more obvious, examples of the type of sampling method being dictated by the nature of the objects sampled occur in testing firecrackers or the like, where the object is indeed destroyed by the sampling procedure.

For this hypothetical example it would appear that sampling without replacement is called for, which implies the hypergeometric distribution as the frequency function for y. But for our purposes we shall limit ourselves to the binomial as the sampling distribution of y, which implies either: (1) sampling with replacement, or (2) use of the binomial as an approximation to the hypergeometric, if we sample without replacement.

In general, the binomial is a "satisfactory" approximation to the hypergeometric if both Y and $(N - Y)$ are "large" relative to sample size. What this last statement implies is that the costs associated with errors due to the approximation should be included in the analysis. Use of the binomial distribution in practice may require more effort than use of the normal distribution in turn, and, as was pointed out previously, the normal may be used to approximate the binomial—quite well in certain circumstances. Again the adequacy of the approximation should be evaluated on a cost basis if possible.

5. Obtaining a best test

Consider adopting the rule: Reject $H: \tilde{P} \geq .5$ in favor of $H_A: \tilde{P} < .5$ whenever y is less than or equal to 6 in a sample of 20. For the sake of brevity, we shall adopt the following notation for the above statement: $(y \leq 6, n = 20)$ will imply rejection of H.

For each value of $\tilde{P}$ in the range of either the hypothesis or the alternative a different binomial distribution holds; hence the probabilities for $y \le 6$ change accordingly. The chain of losses from decision errors are as a result viewed as conditional on the value that $\tilde{P}$ actually assumes in the population. For example, using $(y \le 6, n = 20)$ when $\tilde{P} = .1$ can only result in a type II error: If we reject H (i.e., observe $y \le 6$) we make a correct decision; but if we observe $y > 6$ our rule would cause us to accept H, obviously an error. The probability that in using $(y \le 6, n = 20)$ when $\tilde{P} = .1$ we make this error is just $P(y > 6 \mid \tilde{P} = .1, n = 20) = .0024$ from the binomial tables. Corresponding to this probability is a monetary loss of \$1000 − \$2000(.1) = \$800. On the other hand, under these circumstances if we make the correct decision, and reject H when $\tilde{P} = .1$, we do not suffer loss from a decision error. The probability that we make the correct decision is $P(y \le 6 \mid \tilde{P} = .1, n = 20) = .9976$. The net effect in terms of losses may be viewed as a problem in determining an expected value for a random variable, monetary loss, with possible values \$800 and \$0 and probabilities .0024 and .9976, respectively. The resulting expected loss, \$1.92, is then the amount we may expect to lose, on the average, by using the rule $(y \le 6, n = 20)$ when $\tilde{P} = .1$.

$\tilde{P}$, of course, is not restricted to a value of .1. Indeed, as far as we know, $\tilde{P}$ may be any value between zero and 1. It should be clear at this point that for $\tilde{P} < .5$, the only decision errors accruing to our fixed rule will be type II errors. If $\tilde{P} \ge .5$, then type I errors occur.

Consider the case for $\tilde{P} = .5$. A type I error will occur if we use the rule $(y \le 6, n = 20)$ and observe $y \le 6$. This outcome has probability $P(y \le 6 \mid \tilde{P} = .5, n = 20) = .0577$ of occurring, and corresponding to its occurrence we stand to lose nothing by this error, because .5 is a breakeven point. A "correct" decision, to accept H and $\tilde{P} = .5$ in reality is made whenever $y > 6$, which will happen with probability .9423. The net effect of this particular circumstance is an expected monetary loss of (\$0)(.0577) + (\$0)(.9423) = \$0.

For $\tilde{P} = .6$, and reasoning as above, we obtain an expected loss of (\$200)(.0065) + (\$0)(.9935) = \$1.30. Listed below are identical calculations for other values of $\tilde{P}$.

(a) $E_{\text{loss}}\ (\tilde{P} = .2) = \$52.02$

(b) $E_{\text{loss}}\ (\tilde{P} = .3) = \$156.80$

(c) $E_{\text{loss}}\ (\tilde{P} = .4) = \$150.00$

(d) $E_{\text{loss}}\ (\tilde{P} = .7) = \$0.12$

By classical standards, our best rule (if it exists) will be one which provides minimum losses over all values of $\tilde{P}$. Carrying out the calculations for some other rules, we obtain the entries in Table 7-1. These values are also

Table 7-1. Expected loss calculations for the rules $c = 5$, $c = 6$, $c = 7$, $c = 8$, $c = 9$, $n = 20$; complex hypothesis, complex alternative case

			c									
			5		**6**		**7**		**8**		**9**	
	$\tilde{P}$	**Error loss, $**	**Prob. of loss**	**E_{loss}, $**	**Prob. of loss**	**E_{loss}, $**	**Prob. of loss**	**E_{loss}, $**	**Prob. of loss**	**E_{loss}, $**	**Prob. of loss**	**E_{loss}, $**
Type II	.1	800	.0113	9.04	.0024	1.92	.0004	.32	.0001	.08	$\doteq 0$	$\doteq 0$
	.2	600	.1958	117.48	.0867	52.02	.0321	19.26	.0100	6.00	.0026	1.56
	.3	400	.5836	233.44	.3920	156.80	.2277	91.08	.1133	45.32	.0480	19.20
	.4	200	.8744	174.88	.7500	150.00	.5841	116.82	.4044	80.88	.2447	48.95
Type I	.5	0	.0207	$\doteq 0$	.0577	$\doteq 0$	.1316	$\doteq 0$	.2517	$\doteq 0$	.4119	$\doteq 0$
	.6	200	.0016	.32	.0065	1.30	.0210	2.40	.0565	11.30	.1275	25.50
	.7	400	$\doteq 0$	$\doteq 0$	.0003	.12	.0013	.52	.0051	2.04	.0171	6.84
	.8	600	$\doteq 0$	$\doteq 0$	$\doteq 0$	$\doteq 0$	$\doteq 0$	$\doteq 0$	.0001	.06	.0006	.12

Note: $P(y = c \mid 20, \tilde{P}) = P[y = (20 - c) \mid 20, (1 - \tilde{P})]$, the symmetry of the binomial. Thus $P(y \leq c \mid 20, \tilde{P}) = P[y \geq (20 - c) \mid 20, (1 - \tilde{P})]$. From this relation the above probabilities for $\tilde{P} = .6, .7, .8$ may be determined from the Appendix Tables.

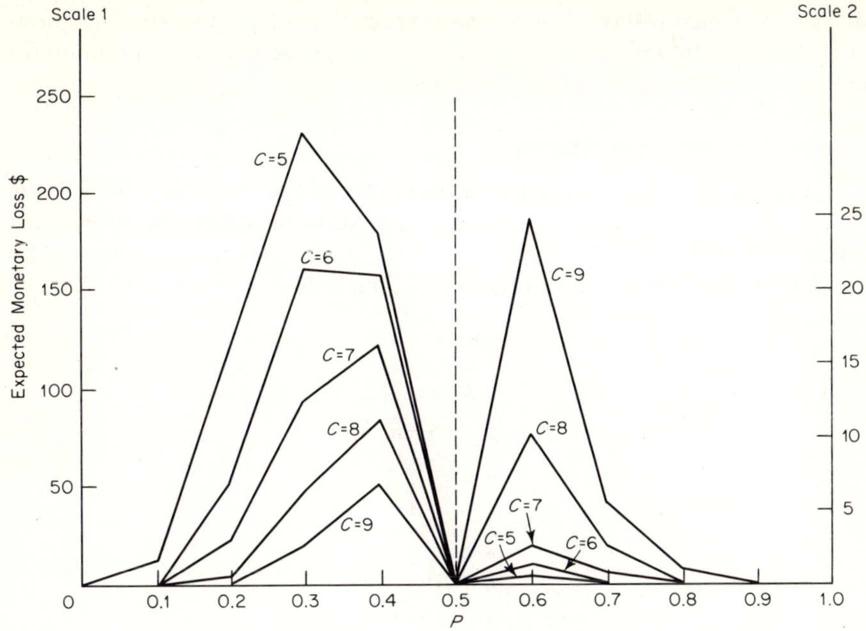

Figure 7-1. Loss characterization.

graphed in Figure 7-1, where we note that no rule provides minimum losses over all values for $\tilde{P}$.†

We should note at this stage of the development that since α errors are *valued* differently in the range of the hypothesis, we cannot proceed to deduce the best rule merely by controlling α and looking for that rule which produces minimum β (weighted) over the alternative. So far we have extended our discussion of classical testing to cover only this case; the modified criterion would call for a rule which produced (weighted) minimum β and α over $\tilde{P} < .5$ and $\tilde{P} \geq .5$, respectively.

With a strict classical definition of the term parameter, we cannot hope to resolve the problem which now faces us. There is no way to choose between alternative rules except by a value judgment or by the introduction of new criteria, because none of the rules in Figure 7-1 meet the classical standard. But if we give credence to a modern interpretation (and I think the distinction implied by the use of *classical* and *modern* is extreme), note that the conditional expected losses associated with each rule can be viewed, again, as random variables which take on their particular values with variations in $\tilde{P}$. Hence we can introduce the notion of expected losses from a given rule over all

† We have computed only for selected values of $\tilde{P}$. For example, $\tilde{P}$ changes by increments of $1/500 = .002$, and actually the graph (and the entire analysis) should involve all 501 possible values for $\tilde{P}$.

values of $\tilde{P}$, which will result in a *single* expected loss from incorrect decisions for each rule. That there must be a rule (or rules) which yields a minimum for such losses will be argued intuitively below.

6. The "bayesian" contribution

Suppose we were able to obtain some idea of the probable values for $\tilde{P}$. That is, from some source we obtain information regarding the probability distribution of $\tilde{P}$, say in the form of a historical frequency distribution. This distribution, the "prior" distribution of $\tilde{P}$, might look as shown in Table 7-2.

Table 2. Distribution of $\tilde{P}$

$\tilde{P}$	$P(\tilde{P})$
.2	.1
.3	.1
.4	.1
.5	.1
.6	.4
.7	.2
	1.0

How can we incorporate such information into our analysis of the best decision rule to follow? Let us not, at this point, raise the issue of the value of such information, but accept the prior distribution without question.

Then each expected loss in Table 7-1 should be interpreted as a conditional loss; e.g., for the rule ($y \le 6$, $n = 20$) the expected loss of \$52.02 is conditional on $\tilde{P}$ taking the value .2, which it does, according to our prior distribution, with probability .1. Hence the (joint) expected loss from using this rule when $\tilde{P} = .2$ is simply $(.1)(.0867)(\$600) = (.1)(\$52.02) = \$5.20$. Proceeding similarly for the remaining possible values of $\tilde{P}$ we obtain

$$\begin{aligned} E_{\text{loss}}(y \le 6, 20) &= (.1)(\$52.02) + (.1)(\$156.80) + (.1)(\$150.00) \\ &\quad + 0 + (.4)(\$1.30) + (.2)(\$.12) \\ &= \$41.10. \end{aligned}$$

Identical calculations for the rules $c = 5$, $c = 7$, $c = 8$, $c = 9$ yield

$$\begin{aligned} E_{\text{loss}}(y \le 5, 20) &= \$52.71, \\ E_{\text{loss}}(y \le 6, 20) &= \$41.10, \\ E_{\text{loss}}(y \le 7, 20) &= \$23.78, \\ E_{\text{loss}}(y \le 8, 20) &= \$18.15, \\ E_{\text{loss}}(y \le 9, 20) &= \$18.54. \end{aligned}$$

The rule ($y \leq 8, 20$), which apparently yields minimum expected loss under our assumed prior distribution, is appropriately called a bayesian decision rule, since we make use of a weighted average of expectations to obtain the ultimate expected losses accruing to our set of possible decision rules, the weights being the probabilities in the prior distribution of $\tilde{P}$. The selection of the rule ($y \leq 8, 20$) as the "best" rule to use is, of course, conditioned on the assumption that our friend's utility for money is linear; hence a minimization of expected monetary loss is appropriate. If this is not the case, an appropriate utility scale must be constructed first; the analysis then proceeds to deduce the rule which yields minimum disutility.

It should be clear that a minimum for expected total loss over alternative rules will exist simply by studying Table 7-1. If more rules were included one would see a complete spectrum of movement from rules where the expected losses from making type I errors are essentially zero with very high expected losses from type II errors (c small) to the other extreme, where the type I and type II expected losses are in juxtaposition. As long as the prior distribution assigns some nonzero probabilities to the occurrence of both error types, a minimum will exist.

7. Sample-size determination

At the outset of our discussion we made two assumptions, regarding sample size and the cost of sampling, which we shall now relax.

If, in fact, sample size is fixed, the cost of sampling is obviously fixed over all decision rules, and so the optimal rule remains unchanged. The necessary change in our analysis would be to add the cost of obtaining the sample to the expected loss from incorrect decisions to obtain an expected total loss. The only situation which realistically portrays the fixed-sample problem would occur when the sample already has been taken—where the costs of sampling have already been incurred.

The question then arises, however, as to how we should treat the sample information. In classical statistics whether or not the sample has already been taken is a moot point. The decision rule would be constructed as if no sample had been taken.

From the modern point of view the sample already taken constitutes additional information which should be incorporated as evidence *before* the optimum rule is determined. In point of fact, all the analysis up to this point of our discussion and the following treatment of sample-size determination should proceed with the distribution which results from a combination of the prior distribution and the sample information. The resultant distribution, formed by applying Bayes' theorem, is known as a *posterior* distribution; we shall give an example of such a computation in Section 9. Thus sample-size determination, *including a decision to sample or not*, is a problem which arises even when a sample has already been taken.

To begin we shall assume that sampling costs are known, and may be represented as a function of n, which for most cases will be an increasing function. Suppose we assume a very simple linear function to represent sampling cost:

$$\text{sampling cost} = a + bn,$$

where a is a fixed cost, incurred regardless of the sample size, and b is the per unit, or per observation, cost. Then we may combine the sampling costs for each n with the expected losses from incorrect decisions for the *optimum* decision rule for each n to determine an expected optimal total loss function over all values of n. What will this function look like?

It can be shown mathematically that the locus of optimum expected losses from decision errors is decreasing over n, so that if sampling were free, the decision maker could reduce his expected losses from decision errors to zero merely by taking a large enough sample. Another algebraic proof gives the following fact: If a rule (c, n) is optimum for some n, then when n is incremented by 1, if the expected loss from errors $E_{\text{loss}}(c, n + 1) \leq E_{\text{loss}}(c, n)$, the rejection point c is optimal for $n + 1$; and if $E_{\text{loss}}(c, n + 1) > E_{\text{loss}}(c, n)$, the optimal rule is $(c + 1, n + 1)$.

Loosely interpreted, the above statement says we may proceed to derive the locus of optimum points by incrementing sample size by 1, testing for optimality, and either incrementing c by 1 or proceeding to increment sample size again by 1. This procedure has no direct bearing on our analysis, but it saves us from completing the set of calculations in Table 7-1 for each n. Moreover, if sampling cost is an increasing function of n and the expected loss from errors is a decreasing function of n, a saddle point will exist on the expected optimal total loss function which is the sum of these two; hence we can be sure there is an optimum n.

The three functions are shown in Figure 7-2 for our example with an assumed sampling cost of \$20 + \$4n. Note at the bottom of the graph that the optimum rules are indicated for all values of n between zero and 20.

The value for $n = 0$ is the expected loss from decision errors when no sample is taken.† How is this loss determined? Before a sample is taken our friend may either reject or not reject the hypothesis solely on the evidence of the prior (or posterior) distribution of $\tilde{P}$. If he rejects H and $\tilde{P} = .5, .6$, or $.7$, he errs. If he "accepts" H and $\tilde{P} = .2, .3$, or $.4$, he errs. The expected losses from such decision errors can be calculated as

$$\begin{aligned} E_{\text{loss}}(\text{accept}) &= (.1)(\$600) + (.1)(\$400) + (.1)(\$200) \\ &\quad + (.1)(\$0) + (.4)(\$0) + (.2)(\$0) \\ &= \$120, \end{aligned}$$

† Or no additional sample if working with a posterior distribution of $\tilde{P}$.

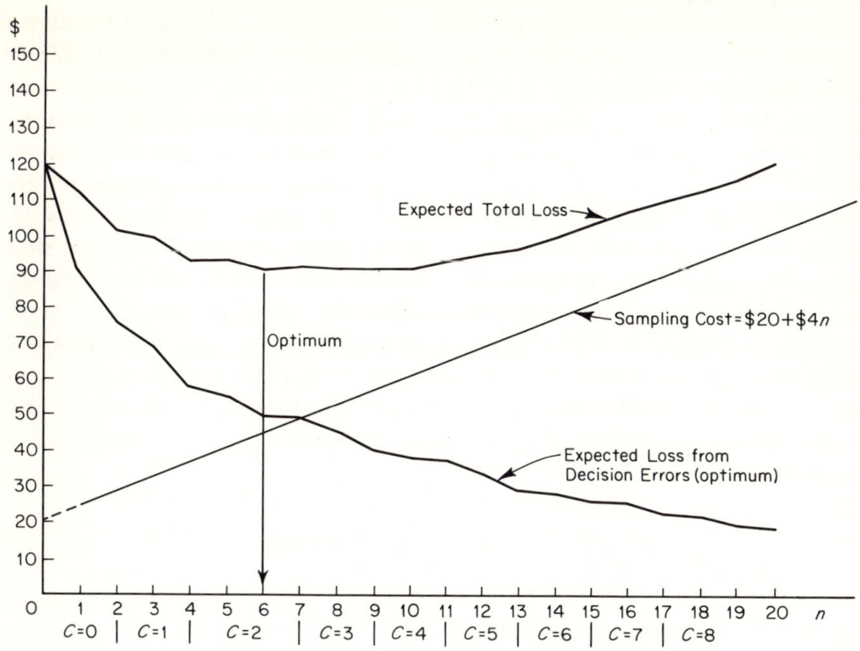

Figure 7-2. Expected total loss and optimum sample size.

whereas

$$E_{\text{loss}}(\text{reject}) = (.1)(\$0) + (.1)(\$0) + (.1)(\$0) + (.4)(\$200) + (.2)(\$400)$$
$$= \$160.$$

If our friend does not sample, the optimum decision to make is to dig up all 500 oysters. He can expect to lose \$120, on the average, from incorrect decisions.

Next we attempt to answer the questions: Should a sample be taken, and if so, how large should it be? The answer is determined by inspecting Figure 7-2. Given the assumed sampling cost function, the optimum n occurs at $n = 6$, where, with the rule ($y \leq 2$), expected total loss amounts to \$89.25. Since this loss is smaller than the \$120 loss when no sample is taken, our friend should take $n = 6$ and reject $H : \tilde{P} \geq .5$ if the number of oysters with pearls in the sample is 2 or less. This, according to our analysis, is the best procedure to follow (again assuming a linear utility-money relation). This cost may also be interpreted as the cost of uncertainty from the decision maker's point of view.

Note that the expected total loss function in Figure 7-2 exhibits strange behavior. For example, from $n = 3$ to $n = 4$ expected total loss decreases, but from $n = 4$ to $n = 5$ losses increase slightly, even though the trend in the function from $n = 0$ to $n = 6$ is downward. This behavior is due to the shape of the expected loss function of decision errors, which exhibits a continuous downward trend but is "kinked." That is to say, at those values of n where a different rule becomes optimum the decrease in the E_{loss} function overshadows the increase in sampling cost; hence the expected total loss function shows a decrease. Until this changeover point is reached, the decrease in E_{loss} may be less than the increase in sampling cost, which causes expected total loss to increase. This behavior is not peculiar to an underlying discrete distribution.

As a final topic in the analysis of optimum sample size, it may not be possible to specify in advance the cost of sampling. An example of this situation would arise if we considered the possibility that our friend could sell the pearls obtained from the sample for \$4.00 each. For a sample of size n the sampling cost would be \$20 + \$4n ($n > 0$), but for every oyster that contained a pearl we could recover \$4 by selling the pearl. According to the prior distribution of $\tilde{P}$, we could expect 20 per cent of the sample to contain oysters 10 per cent of the time, etc., so that for any $n > 0$ our expected revenue from the sale of pearls would be

$$\begin{aligned} E_{\text{revenue}} &= \$4[(.1)(.2)n + (.1)(.3)n + (.1)(.4)n + (.1)(.5)n \\ &\quad + (.4)(.6)n + (.2)(.7)n] \\ &= \$4(.42)n \\ &= \$1.68n. \end{aligned}$$

Thus, under these circumstances our sampling cost, now *expected* sampling cost, would be

$$\begin{aligned} E_{\text{sampling cost}} &= \$20 + \$4n - \$1.68n \\ &= \$20 + \$2.32n. \end{aligned}$$

This function would replace the sampling cost function of Figure 7-2, a new optimum n and decision rule being determined.

8. Further generalizations

Although we have concentrated the development of the basic ideas of modern decision theory on an example which utilizes a discrete distribution, the methods can equally well be applied to the continuous case, for example, where the sampling distribution is normal.† The importance of the normal distribution to both classical and modern statistical decision making primarily derives from the fundamental central limit theorem, which demonstrates that

† For an illustration the reader is again referred to Schlaifer, *op. cit.* (pp. 269–315).

many sampling distributions tend to the normal distribution for large sample sizes, including the binomial and hypergeometric of our example.

Also, although the test of our example is one-sided (rejection for very small values of the sample statistic), problems may arise which call for a one-sided test for the upper tail of the sampling distribution, or two-sided tests where both very large and very small values of the sample statistic may demonstrate inconsistency with a main hypothesis. Extensions of the basic methodology to these situations should be obvious.

9. Final remarks

There remain some rather serious questions regarding the practicality and limitations of the methodology presented, as well as additional conceptual problems which have not as yet been answered to the satisfaction of many. We shall only treat the more significant of these here.

At the beginning of our introduction to the material we mentioned the possibility that actual cost figures for decision errors might not be obtainable. Certainly these costs play a major role in the assessment of decision rules for both the classical and modern approaches, but their absence would not detract from the conceptual advantage of the modern interpretation. As to practical advantages of the modern method when cost figures are not available, one might assume, as is many times in practice, that type I and type II errors are equally distasteful; thus our analysis would proceed merely by minimizing the *probabilities* of making errors. Or, as many representatives of the modern school contend, one should proceed to "determine" the costs of errors in terms of utilities, by presenting the decision maker with various pairs of alternative "bets" for which he will judge a preference for one over the other, or indifference.† In this way a utility scale depicting the decision maker's preference patterns with respect to risk may be constructed.‡

Two objections to this latter suggestion which deserve consideration are the following: First, suppose the decision maker has no vested interest in his decision. That is to say, it is conceivable that a research worker would be completely divorced from any personal attachment to some statistical test he wished to perform. Then what sort of utility scale could be constructed for the costs of decision errors?

Second, it is also inconceivable that anyone could make *precise* judgments between these so-called alternative "bets," which must be assumed to obtain the unique utility scale required in lieu of monetary costs. How do these errors in judgment affect the modern approach?

Another practical disadvantage which presently permeates the modern approach cannot, I think, be classified as a criticism. If it is a criticism, it

† See Schlaifer, *op. cit.*, Chap. 2.

‡ It may not be long before every statistician has a couch in his office to aid in analyzing his "patients."

applies to all sciences in their infancy. The point is that many important classical tests concern more than one parameter simultaneously. As yet little research has appeared dealing with conceptual bridges between the classical and modern methods of inference for these problems.

Finally, and perhaps most important, is the matter of prior information. In the classical view it is relatively easy to assess the relative importance of objective (sample) information. That is, the results of a sample of 500 are more reliable than those from a sample of 50, given random sampling, etc. But how reliable are the subjective estimates of a businessman, for example, about the probability distribution of a "parameter"? And, how does one assess this reliability? And then, if reliability cannot be measured, how can we be sure that the results of our analyses have any meaning whatsoever?†

To point out specifically how sample information may be readily subsumed in the analysis of Sections 6 and 7, consider our example again, but suppose that in addition to the prior distribution a sample of size 10 had already been taken, and that 4 of the 10 oysters were found to contain pearls. Intuitively, we might reason that this observation tends to refute $H: \tilde{P} \geq .5$ in favor of $H_A: \tilde{P} < .5$. The question is: How much "weight" should be placed on this information, or, in other words, how should we modify the prior distribution to reflect these results? A moment's reflection should suggest that we are actually asking a question of the form: Given the observed result, what are the appropriate probabilities that it was produced by the various values of $\tilde{P}$ under consideration? This question is precisely the sort for which Bayes' theorem yields answers. The required inputs for an answer are the prior probabilities for the values of $\tilde{P}$ and conditional probabilities for the sample outcome given these values. Calculations for the hypothetical situation are given in Table 7-3, where we note that the prior distribution has been adjusted

Table 7-3. Calculation of a posterior distribution

Value of $\tilde{P}$	$P(\tilde{P})$ "prior"	$P(y = 4/n = 10, \tilde{P})$	Joint prob. $P(y = 4/n = 10, P) \times [P(\tilde{P})]$	Revised prob. $P(\tilde{P})$ "posterior"
.2	.1	.0881	.00881	.070
.3	.1	.2001	.02001	.158
.4	.1	.2508	.02508	.199
.5	.1	.2051	.02051	.162
.6	.4	.1114	.04456	.353
.7	.2	.0367	.00734	.058
			$\Sigma = .12631$	1.000

† This problem is not unique to the modern approach. Its classical counterparts are assumptions about the underlying distribution being, e.g., normal, etc.

to show the effects of the sample result; accordingly, the "likelihood" that $\tilde{P}$ has value .2, .6, or .7 has been adjusted downward, and the converse for the likelihoods of $\tilde{P} = .3$, .4, or .5. As mentioned previously, the discovery of a "best" decision rule would now follow, using the posterior distribution for $\tilde{P}$.

Still, the calculations just carried out depend on a statement of prior probabilities, and these, in turn, may be of a subjective rather than objective nature. Obviously our complete analysis of the best decision rule is based upon the prior distribution to at least some degree.†

Two points are to be made concerning this last statement. If our analysis purports to be a model of the decision-making process under uncertainty, then these subjective probabilities are simply quantitative representations of a decision maker's subjective judgment about a given random outcome. Thus, given that such judgments are made, the prior distribution is nothing more than a formal statement of them. Finally, if the subjective judgment we are speaking of is one of "no information," then its counterpart for us is a rectangular prior distribution of the parameter in question, e.g., an assumption that every possible value of the parameter is equally likely.

Finally, the classical confidence interval has its modern counterpart, defined via an analysis similar to that used for developing the binomial and Poisson intervals (Section 2, Chapter 6), and based on a hypothesis test of the form $H: \mu = \mu_0$ vs. $H_A: \mu \neq \mu_0$. Thus α-probability may be controlled and for a given sample result (for a given sample size) we may compare the rejection limits for various assumed values of μ_0 to produce the interval.

Extensions of the modern analysis, including the specification of continuous prior and sampling distributions are possible, but not in all cases are they easy to accomplish. Exceptions include the normal prior–normal sampling distribution and uniform‡ (rectangular) prior cases.

† If we carried on the process of taking samples, and used Bayes' theorem to produce new posterior distributions from old ones by a sequential incorporation of sample information, we would, under ideal circumstances, eventually arrive at a posterior distribution in which a single value of the parameter in question had probability 1 of being the actual parameter no matter what the prior probability distribution (as long as it allowed all values of $\tilde{P}$ to be possible). We might liken this phenomenon to that of consistent estimation (cf. Chapter 5).

‡ Where each value of the random variable possesses the same density.

8

Recapitulation

"It's not whether you win or lose,
it's how you play the game"

the locker room door

Although the above quotation may express the sentiments of a very moral football coach (or one who does not have to answer to his alumni), it does not serve to argue the practical advantages of statistical decision-making over "intuition," "judgment," "experience," or whatever apparently non-systematic alternative means for making decisions under uncertainty we might consider.

What we have outlined in this book is a framework for decision-making under uncertainty which (1) may be a formal statement of how decisions under uncertainty are already made in practice, but (2) more likely is a statement of how decisions should be made, given that in any particular application it is appropriate to apply the methodology, i.e., the prerequisite assumptions are met. The major contribution of the method, from the decision maker's point of view, is in the systematic treatment and subsequent control of decision errors (in a probabilistic sense), leading to better decisions.

Although our development has proceeded from a primarily theoretical bent and has, moreover, been given from a limited view of specific application, it is reasserted here that the methodology may be extended easily, in the sense of conceptual ease. Thus to consider tests of hypotheses about population means only (including proportions) does not restrict the area of application for the technique as a whole. One need only consult more exhaustive texts on hypothesis testing for a larger catalogue.

The progression of subject matter from the outset has been directed toward the expositions of Chapters 6 and 7, although some of the background material may bear no apparent relevance to what is *actually* done in these chapters. The reason for this is simple, in lieu of the need to present topics for purposes of building the reader to a required level of competence and sophistication: although the decision-making aspects of applied statistics are of major importance, they are treated via a limited number of hypothetical—indeed, the simplest—cases. So, for example, we discuss the estimation of the population variance and use the results for purposes of testing a hypothesis about the population mean, but do not consider specifically hypothesis tests about this variance. Or, the ideas of joint frequency are not exploited to any great extent in the presentation even though they could have been had we considered, for example, the maximum likelihood method of estimation in some detail. As it is, we merely allude to the joint distribution of the random variables used to describe the sampling process.

What is (or should be) available from the subject matter discussed and the context in which it is considered is an understanding of the logical structure which supports statistical decision-making, with the expectation that the reader may be a more intelligent user of statistical methods as they apply to choice under uncertainty, via first a proper assessment of the appropriate probability model (sampling distribution) and the quality of subsequent judgments (inferences) derived from it.

Tables

Table 1.† Cumulative binomial distributions

$$P(y \geq k \mid n, p) = \sum_{y=k}^{n} \binom{n}{k} p^k q^{n-k}$$

$n = 5$	*p*				
k	**.1**	**.2**	**.3**	**.4**	**.5**
1	.4095	.6723	.8319	.9222	.9688
2	.0815	.2627	.4718	.6630	.8125
3	.0086	.0579	.1631	.3174	.5000
4	.0005	.0067	.0308	.0870	.1875
5	.0000	.0003	.0024	.0102	.0312

$n = 10$	*p*				
k	**.1**	**.2**	**.3**	**.4**	**.5**
1	.6513	.8926	.9717	.9939	.9990
2	.2639	.6242	.8507	.9536	.9893
3	.0702	.3222	.6172	.8327	.9453
4	.0128	.1209	.3504	.6177	.8281
5	.0016	.0328	.1503	.3669	.6230
6	.0001	.0064	.0473	.1662	.3770
7	.0000	.0009	.0106	.0548	.1719
8	.0000	.0001	.0016	.0123	.0547
9	.0000	.0000	.0001	.0017	.0107
10	.0000	.0000	.0000	.0001	.0010

† Extracted with permission from *Tables of the Binomial Probability Distribution* National Bureau of Standards, Applied Mathematics Series 6 (Washington, D.C., 1950).

Table 1 (*continued*)

$n = 15$	p				
k	.1	.2	.3	.4	.5
1	.7941	.9648	.9953	.9995	1.0000
2	.4510	.8329	.9647	.9948	.9995
3	.1841	.6020	.8732	.9729	.9963
4	.0556	.3518	.7031	.9095	.9824
5	.0127	.1642	.4845	.7827	.9408
6	.0022	.0611	.2784	.5968	.8491
7	.0003	.0181	.1311	.3902	.6964
8	.0000	.0042	.0500	.2131	.5000
9	...	.0008	.0152	.0950	.3036
10	...	.0001	.0037	.0338	.1509
11	...	.0000	.0007	.0093	.0592
12	...	...	.0001	.0019	.0176
13	...	...	.0000	.0003	.0037
14	...	...	...	.0000	.0005
15	...	...	...	.0000	.0000

$n = 20$	p				
k	.1	.2	.3	.4	.5
1	.8784	.9885	.9992	1.0000	1.0000
2	.6083	.9308	.9924	.9995	1.0000
3	.3231	.7939	.9645	.9964	.9998
4	.1330	.5886	.8929	.9840	.9987
5	.0432	.3704	.7625	.9490	.9941
6	.0113	.1958	.5836	.8744	.9793
7	.0024	.0867	.3920	.7500	.9423
8	.0004	.0321	.2277	.5841	.8684
9	.0001	.0100	.1133	.4044	.7483
10	.0000	.0026	.0480	.2447	.5881
11	....	.0006	.0171	.1275	.4119
12	....	.0001	.0051	.0565	.2517
13	....	.0000	.0013	.0210	.1316
14	....	....	.0003	.0065	.0577
15	....	....	.0000	.0016	.0207
16	....	....	....	.0003	.0059
17	....	....	....	.0000	.0013
18	....	....	....	....	.0002
19	....	....	....	....	.0000
20	....	....	....	....	.0000

Table 1 (*continued*)

$n = 30$	p				
k	.1	.2	.3	.4	.5
1	.9576	.9988	1.0000	1.0000	1.0000
2	.8163	.9895	.9997	1.0000	1.0000
3	.5886	.9558	.9979	1.0000	1.0000
4	.3526	.8773	.9907	.9997	1.0000
5	.1755	.7448	.9698	.9985	1.0000
6	.0732	.5725	.9234	.9943	.9998
7	.0258	.3930	.8405	.9828	.9993
8	.0078	.2392	.7186	.9565	.9974
9	.0020	.1287	.5685	.9060	.9919
10	.0005	.0611	.4112	.8237	.9786
11	.0001	.0256	.2696	.7085	.9506
12	.0000	.0095	.1593	.5689	.8998
13	....	.0031	.0845	.4215	.8192
14	....	.0009	.0401	.2855	.7077
15	....	.0002	.0169	.1754	.5722
16	....	.0001	.0064	.0971	.4278
17	....	.0000	.0021	.0481	.2923
18	....	....	.0006	.0212	.1808
19	....	....	.0002	.0083	.1002
20	....	....	.0000	.0029	.0494
21	....	....	....	.0009	.0214
22	....	....	....	.0002	.0081
23	....	....	....	.0000	.0026
24	....	....	....	....	.0007
25	....	....	....	....	.0001
26	....	....	....	....	.0000
27	....	....	....	....	....
28	....	....	....	....	....
29	....	....	....	....	....
30	....	....	....	....	....

Table 2.† Normal distribution function $\Phi(z) = \frac{1}{\sqrt{2\pi}} \int_{-\infty}^{z} e^{-(x^2/2)}\, dx$ for $0.00 \leq z \leq 4.99$

z	.00	.01	.02	.03	.04	.05	.06	.07	.08	.09
.0	.5000	.5040	.5080	.5120	.5160	.5199	.5239	.5279	.5319	.5359
.1	.5398	.5438	.5478	.5517	.5557	.5596	.5636	.5675	.5714	.5753
.2	.5793	.5832	.5871	.5910	.5948	.5987	.6026	.6064	.6103	.6141
.3	.6179	.6217	.6255	.6293	.6331	.6368	.6406	.6443	.6480	.6517
.4	.6554	.6591	.6628	.6664	.6700	.6736	.6772	.6808	.6844	.6879
.5	.6915	.6950	.6985	.7019	.7054	.7088	.7123	.7157	.7190	.7224
.6	.7257	.7291	.7324	.7357	.7389	.7422	.7454	.7486	.7517	.7549
.7	.7580	.7611	.7642	.7673	.7703	.7734	.7764	.7794	.7823	.7852
.8	.7881	.7910	.7939	.7967	.7995	.8023	.8051	.8078	.8106	.8133
.9	.8159	.8186	.8212	.8238	.8264	.8289	.8315	.8340	.8365	.8389
1.0	.8413	.8438	.8461	.8485	.8508	.8531	.8554	.8577	.8599	.8621
1.1	.8643	.8665	.8686	.8708	.8729	.8749	.8770	.8790	.8810	.8830
1.2	.8849	.8869	.8888	.8907	.8924	.8944	.8962	.8980	.8997	.90147
1.3	.90320	.90490	.90658	.90824	.90988	.91149	.91309	.91466	.91621	.91774
1.4	.91924	.92073	.92220	.92364	.92507	.92647	.92785	.92922	.93056	.93189
1.5	.93319	.93448	.93574	.93699	.93822	.93943	.94062	.94179	.94295	.94408
1.6	.94520	.94630	.94738	.94845	.94950	.95053	.95154	.95254	.95352	.95449
1.7	.95543	.95637	.95728	.95818	.95907	.95994	.96080	.96164	.96246	.96327
1.8	.96407	.96485	.96562	.96638	.96712	.96784	.96856	.96926	.96995	.97062
1.9	.97128	.97193	.97257	.97320	.97381	.97441	.97500	.97558	.97615	.97670
2.0	.97725	.97778	.97831	.97882	.97932	.97982	.98030	.98077	.98124	.98169
2.1	.98214	.98257	.98300	.98341	.98382	.98422	.98461	.98500	.98537	.98574
2.2	.98610	.98645	.98679	.98713	.98745	.98778	.98809	.98840	.98870	.98899
2.3	.98928	.98956	.98983	$.9^2$ 0097	$.9^2$ 0358	$.9^2$ 0613	$.9^2$ 0863	$.9^2$ 1106	$.9^2$ 1344	$.9^2$ 1576
2.4	$.9^2$ 1802	$.9^2$ 2024	$.9^2$ 2240	$.9^2$ 2451	$.9^2$ 2451	$.9^2$ 2857	$.9^2$ 3053	$.9^2$ 3244	$.9^2$ 3431	$.9^2$ 3613

2.5	$.9^2$ 3790	$.9^2$ 3963	$.9^2$ 4132	$.9^2$ 4297	$.9^2$ 4457	$.9^2$ 4614	$.9^2$ 4766	$.9^2$ 4915	$.9^2$ 5060	$.9^2$ 5201
2.6	$.9^2$ 5339	$.9^2$ 5473	$.9^2$ 5604	$.9^2$ 5731	$.9^2$ 5855	$.9^2$ 5975	$.9^2$ 6093	$.9^2$ 6207	$.9^2$ 6319	$.9^2$ 6427
2.7	$.9^2$ 6533	$.9^2$ 6636	$.9^2$ 6736	$.9^2$ 6833	$.9^2$ 6928	$.9^2$ 7020	$.9^2$ 7110	$.9^2$ 7197	$.9^2$ 7282	$.9^2$ 7365
2.8	$.9^2$ 7445	$.9^2$ 7523	$.9^2$ 7599	$.9^2$ 7673	$.9^2$ 7744	$.9^2$ 7814	$.9^2$ 7882	$.9^2$ 7948	$.9^2$ 8012	$.9^2$ 8074
2.9	$.9^2$ 8134	$.9^2$ 8193	$.9^2$ 8250	$.9^2$ 8305	$.9^2$ 8359	$.9^2$ 8411	$.9^2$ 8462	$.9^2$ 8511	$.9^2$ 8559	$.9^2$ 8605
3.0	$.9^2$ 8650	$.9^2$ 8694	$.9^2$ 8736	$.9^2$ 8777	$.9^2$ 8817	$.9^2$ 8856	$.9^2$ 8893	$.9^2$ 8930	$.9^2$ 8965	$.9^2$ 8999
3.1	$.9^3$ 0324	$.9^3$ 0646	$.9^3$ 0957	$.9^3$ 1260	$.9^3$ 1553	$.9^3$ 1836	$.9^3$ 2112	$.9^3$ 2378	$.9^3$ 2636	$.9^3$ 2886
3.2	$.9^3$ 3129	$.9^3$ 3363	$.9^3$ 3590	$.9^3$ 3810	$.9^3$ 4024	$.9^3$ 4230	$.9^3$ 4429	$.9^3$ 4623	$.9^3$ 4810	$.9^3$ 4991
3.3	$.9^3$ 5166	$.9^3$ 5335	$.9^3$ 5499	$.9^3$ 5658	$.9^3$ 5811	$.9^3$ 5959	$.9^3$ 6103	$.9^3$ 6242	$.9^3$ 6376	$.9^3$ 6505
3.4	$.9^3$ 6631	$.9^3$ 6752	$.9^3$ 6869	$.9^3$ 6982	$.9^3$ 7091	$.9^3$ 7197	$.9^3$ 7299	$.9^3$ 7398	$.9^3$ 7493	$.9^3$ 7585
3.5	$.9^3$ 7674	$.9^3$ 7759	$.9^3$ 7842	$.9^3$ 7922	$.9^3$ 7999	$.9^3$ 8074	$.9^3$ 8146	$.9^3$ 8215	$.9^3$ 8282	$.9^3$ 8347
3.6	$.9^3$ 8409	$.9^3$ 8469	$.9^3$ 8527	$.9^3$ 8583	$.9^3$ 8637	$.9^3$ 8689	$.9^3$ 8739	$.9^3$ 8787	$.9^3$ 8834	$.9^3$ 8879
3.7	$.9^3$ 8922	$.9^3$ 8964	$.9^4$ 0039	$.9^4$ 0426	$.9^4$ 0799	$.9^4$ 1158	$.9^4$ 1504	$.9^4$ 1838	$.9^4$ 2159	$.9^4$ 2468
3.8	$.9^4$ 2765	$.9^4$ 3052	$.9^4$ 3327	$.9^4$ 3593	$.9^4$ 3848	$.9^4$ 4094	$.9^4$ 4331	$.9^4$ 4558	$.9^4$ 4777	$.9^4$ 4988
3.9	$.9^4$ 5190	$.9^4$ 5385	$.9^4$ 5573	$.9^4$ 5753	$.9^4$ 5926	$.9^4$ 6092	$.9^4$ 6253	$.9^4$ 6406	$.9^4$ 6554	$.9^4$ 6696
4.0	$.9^4$ 6833	$.9^4$ 6964	$.9^4$ 7090	$.9^4$ 7211	$.9^4$ 7327	$.9^4$ 7439	$.9^4$ 7546	$.9^4$ 7649	$.9^4$ 7748	$.9^4$ 7843
4.1	$.9^4$ 7934	$.9^4$ 8022	$.9^4$ 8106	$.9^4$ 8186	$.9^4$ 8263	$.9^4$ 8338	$.9^4$ 8409	$.9^4$ 8477	$.9^4$ 8542	$.9^4$ 8605
4.2	$.9^4$ 8665	$.9^4$ 8723	$.9^4$ 8778	$.9^4$ 8832	$.9^4$ 8882	$.9^4$ 8931	$.9^4$ 8978	$.9^5$ 0226	$.9^5$ 0655	$.9^5$ 1066
4.3	$.9^5$ 1460	$.9^5$ 1837	$.9^5$ 2199	$.9^5$ 2545	$.9^5$ 2876	$.9^5$ 3193	$.9^5$ 3497	$.9^5$ 3788	$.9^5$ 4066	$.9^5$ 4332
4.4	$.9^5$ 4587	$.9^5$ 4831	$.9^5$ 5065	$.9^5$ 5288	$.9^5$ 5502	$.9^5$ 5706	$.9^5$ 5902	$.9^5$ 6089	$.9^5$ 6268	$.9^5$ 6439
4.5	$.9^5$ 6602	$.9^5$ 6759	$.9^5$ 6908	$.9^5$ 7051	$.9^5$ 7187	$.9^5$ 7318	$.9^5$ 7442	$.9^5$ 7561	$.9^5$ 7675	$.9^5$ 7784
4.6	$.9^5$ 7888	$.9^5$ 7987	$.9^5$ 8081	$.9^5$ 8172	$.9^5$ 8258	$.9^5$ 8340	$.9^5$ 8419	$.9^5$ 8494	$.9^5$ 8566	$.9^5$ 8634
4.7	$.9^5$ 8699	$.9^5$ 8761	$.9^5$ 8821	$.9^5$ 8877	$.9^5$ 8931	$.9^5$ 8983	$.9^6$ 0320	$.9^6$ 0789	$.9^6$ 1235	$.9^6$ 1661
4.8	$.9^6$ 2067	$.9^6$ 2453	$.9^6$ 2822	$.9^6$ 3173	$.9^6$ 3508	$.9^6$ 3827	$.9^6$ 4131	$.9^6$ 4420	$.9^6$ 4696	$.9^6$ 4958
4.9	$.9^6$ 5208	$.9^6$ 5446	$.9^6$ 5673	$.9^6$ 5889	$.9^6$ 6094	$.9^6$ 6289	$.9^6$ 6475	$.9^6$ 6652	$.9^6$ 6821	$.9^6$ 6981

Example: $\Phi(3.57) = .9^3\,8215 = 0.9997215$.

† From *Statistical Tables and Formulas*, by A. Hald (Wiley, 1952), with permission of the publisher.

Table 3.† Percentage points of the t distribution

ν	$Q = 0.4$	0.25	0.05	0.025	0.005	0.0025	0.001
	$2Q = 0.8$	0.5	0.1	0.05	0.01	0.005	0.002
1	0.325	1.000	6.314	12.706	63.657	127.32	318.31
2	.289	0.816	2.920	4.303	9.925	14.089	22.326
3	.277	.765	2.353	3.182	5.841	7.453	10.213
4	.271	.741	2.132	2.776	4.604	5.598	7.173
5	0.267	0.727	2.015	2.571	4.032	4.773	5.893
6	.265	.718	1.943	2.447	3.707	4.317	5.208
7	.263	.711	1.895	2.365	3.499	4.029	4.785
8	.262	.706	1.860	2.306	3.355	3.833	4.501
9	.261	.703	1.833	2.262	3.250	3.690	4.297
10	0.260	0.700	1.812	2.228	3.169	3.581	4.144
11	.260	.697	1.796	2.201	3.106	3.497	4.025
12	.259	.695	1.782	2.179	3.055	3.428	3.930
13	.259	.694	1.771	2.160	3.012	3.372	3.852
14	.258	.692	1.761	2.145	2.977	3.326	3.787
15	0.258	0.691	1.753	2.131	2.947	3.286	3.733
16	.258	.690	1.746	2.120	2.921	3.252	3.686
17	.257	.689	1.740	2.110	2.898	3.222	3.646
18	.257	.688	1.734	2.101	2.878	3.197	3.610
19	.257	.688	1.729	2.093	2.861	3.174	3.579
20	0.257	0.687	1.725	2.086	2.845	3.153	3.552
21	.257	.686	1.721	2.080	2.831	3.135	3.527
22	.256	.686	1.717	2.074	2.819	3.119	3.505
23	.256	.685	1.714	2.069	2.807	3.104	3.485
24	.256	.685	1.711	2.064	2.797	3.091	3.467
25	0.256	0.684	1.708	2.060	2.787	3.078	3.450
26	.256	.684	1.706	2.056	2.779	3.067	3.435
27	.256	.684	1.703	2.052	2.771	3.057	3.421
28	.256	.683	1.701	2.048	2.763	3.047	3.408
29	.256	.683	1.699	2.045	2.756	3.038	3.396
30	0.256	0.683	1.697	2.042	2.750	3.030	4.385
40	.255	.681	1.684	2.021	2.704	2.971	3.307
60	.254	.679	1.671	2.000	2.660	2.915	3.232
120	.254	.677	1.658	1.980	2.617	2.860	3.160
∞	.253	.674	1.645	1.960	2.576	2.807	3.090

$Q = 1 - P(t \mid \nu)$ is the upper-tail area of the distribution for ν degrees of freedom, appropriate for use in a single-tail test. For a two-tail test, $2Q$ must be used.

† From *Biometrika Tables for Statisticians*, Vol. I, edited by E. S. Pearson and H. O. Hartley (Cambridge, 1958), with permission of the Biometrika Trustees.

Index

W

Y